SAFETY AND SECURITY
OF
COMMERCIAL
SPENT NUCLEAR FUEL STORAGE

Public Report

Committee on the Safety and Security of Commercial Spent Nuclear Fuel Storage

Board on Radioactive Waste Management

Division on Earth and Life Studies

NATIONAL RESEARCH COUNCIL
OF THE NATIONAL ACADEMIES

THE NATIONAL ACADEMIES PRESS
Washington, D.C.
www.nap.edu

THE NATIONAL ACADEMIES PRESS 500 Fifth Street, N.W. Washington, DC 20001

NOTICE: The project that is the subject of this report was approved by the Governing Board of the National Research Council, whose members are drawn from the councils of the National Academy of Sciences, the National Academy of Engineering, and the Institute of Medicine. The members of the committee responsible for the report were chosen for their special competences and with regard for appropriate balance.

This study was supported by grant number NRC-04-04-067 between the National Academy of Sciences and the U.S. Nuclear Regulatory Commission. Any opinions, findings, conclusions, or recommendations expressed in this publication are those of the author(s) and do not necessarily reflect the views of the organizations or agencies that provided support for the project.

International Standard Book Number 0-309-09647-2
Library of Congress Control Number 2005926244

Additional copies of this report are available from the National Academies Press, 500 Fifth Street, N.W., Lockbox 285, Washington, DC 20055; (800) 624-6242 or (202) 334-3313 (in the Washington metropolitan area); Internet, http://www.nap.edu

Printed in the United States of America.

THE NATIONAL ACADEMIES

Advisers to the Nation on Science, Engineering, and Medicine

The **National Academy of Sciences** is a private, nonprofit, self-perpetuating society of distinguished scholars engaged in scientific and engineering research, dedicated to the furtherance of science and technology and to their use for the general welfare. Upon the authority of the charter granted to it by the Congress in 1863, the Academy has a mandate that requires it to advise the federal government on scientific and technical matters. Dr. Ralph J. Cicerone is president of the National Academy of Sciences.

The **National Academy of Engineering** was established in 1964, under the charter of the National Academy of Sciences, as a parallel organization of outstanding engineers. It is autonomous in its administration and in the selection of its members, sharing with the National Academy of Sciences the responsibility for advising the federal government. The National Academy of Engineering also sponsors engineering programs aimed at meeting national needs, encourages education and research, and recognizes the superior achievements of engineers. Dr. Wm. A. Wulf is president of the National Academy of Engineering.

The **Institute of Medicine** was established in 1970 by the National Academy of Sciences to secure the services of eminent members of appropriate professions in the examination of policy matters pertaining to the health of the public. The Institute acts under the responsibility given to the National Academy of Sciences by its congressional charter to be an adviser to the federal government and, upon its own initiative, to identify issues of medical care, research, and education. Dr. Harvey V. Fineberg is president of the Institute of Medicine.

The **National Research Council** was organized by the National Academy of Sciences in 1916 to associate the broad community of science and technology with the Academy's purposes of furthering knowledge and advising the federal government. Functioning in accordance with general policies determined by the Academy, the Council has become the principal operating agency of both the National Academy of Sciences and the National Academy of Engineering in providing services to the government, the public, and the scientific and engineering communities. The Council is administered jointly by both Academies and the Institute of Medicine. Dr. Ralph J. Cicerone and Dr. Wm. A. Wulf are chair and vice chair, respectively, of the National Research Council

www.national-academies.org

COMMITTEE ON THE SAFETY AND SECURITY OF COMMERCIAL SPENT NUCLEAR FUEL STORAGE

LOUIS J. LANZEROTTI, *Chair*, New Jersey Institute of Technology, Newark, and Lucent Technologies, Murray Hill
CARL A. ALEXANDER, Battelle Memorial Institute, Columbus, Ohio
ROBERT M. BERNERO, U.S. Nuclear Regulatory Commission (retired), Gaithersburg, Maryland
M. QUINN BREWSTER, University of Illinois, Urbana-Champaign
GREGORY R. CHOPPIN, Florida State University, Tallahassee
NANCY J. COOKE, Arizona State University, Mesa
LOUIS ANTHONY COX, Jr.,[1] Cox Associates, Inc., Denver, Colorado
GORDON R. JOHNSON, Network Computing Services, Minneapolis, Minnesota
ROBERT P. KENNEDY, RPK Structural Mechanics Consulting, Escondido, California
KENNETH K. KUO, Pennsylvania State University, University Park
RICHARD T. LAHEY, Jr., Rensselaer Polytechnic Institute, Troy, New York
KATHLEEN R. MEYER, Keystone Scientific, Inc., Fort Collins, Colorado
FREDERICK J. MOODY, GE Nuclear Energy (retired), Murphys, California
TIMOTHY R. NEAL, Los Alamos National Laboratory, Los Alamos, New Mexico
JOHN WREATHALL,[1] John Wreathall & Company, Inc., Dublin, Ohio
LORING A. WYLLIE, Jr., Degenkolb Engineers, San Francisco, California
PETER D. ZIMMERMAN, King's College London, United Kingdom

Staff

KEVIN D. CROWLEY, Study Director
BARBARA PASTINA, Senior Program Officer
MICAH D. LOWENTHAL, Senior Program Officer
ELISABETH A. REESE, Program Officer
DARLA THOMPSON, Research Associate
TONI G. GREENLEAF, Administrative Associate

[1] Drs. Cox and Wreathall resigned from the committee on February 26 and March 17, 2004, respectively.

BOARD ON RADIOACTIVE WASTE MANAGEMENT

[1] Dr. Meserve did not participate in the oversight of this study.

ACKNOWLEDGMENTS

This study would not have been possible without the help of several organizations and individuals who were called upon for information and advice. The committee would like to acknowledge especially the following organizations and individuals for their help:

- Congressional staff members Kevin Cook, Terry Tyborowski, and Jeanne Wilson (retired) for their guidance on the study task.
- Nuclear Regulatory Commission staff Farouk Eltawila, who served as the primary liaison for this study, and Charles Tinkler and Francis (Skip) Young for their support of the committee's information-gathering activities.
- Department of Homeland Security staff member Jon MacLaren, who also served as a liaison to the committee.
- Steve Kraft and John Vincent (deceased) of the Nuclear Energy Institute and staff of Energy Resources International for providing information about spent fuel storage practices in industry.
- ENTERGY Corp., Exelon Corp, and Arizona Public Service Corp. staff for organizing tours of the Braidwood, Dresden, Indian Point, and Palo Verde nuclear generating stations.
- German organizations and individuals who helped organize a tour of spent fuel storage facilities in Germany. These organizations and individuals are explicitly acknowledged in Appendix C.
- Speakers (see Appendix A) and participants at committee meetings as well as those who sent written comments for providing their knowledge and perspectives on this important matter.

This report has been reviewed in draft form by individuals chosen for their diverse perspectives and technical expertise, in accordance with procedures approved by the National Research Council's Report Review Committee. The purpose of this independent review is to provide candid and critical comments that will assist the institution in making its published report as sound as possible and to ensure that the report meets institutional standards for objectivity, evidence, and responsiveness to the study charge. The content of the review comments and draft manuscript remain confidential to protect the integrity of the deliberative process. We wish to thank the following individuals for their review of this report:

John F. Ahearne, Sigma Xi and Duke University
Romesh C. Batra, Virginia Polytechnic Institute and State University
Robert J. Budnitz, Lawrence Livermore National Laboratory
Philip R. Clark, GPU Nuclear Corporation (retired)
Richard L. Garwin, IBM Thomas J. Watson Research Center
Roger L. Hagengruber, The University of New Mexico
Darleane C. Hoffman, E.O. Lawrence Berkeley National Laboratory
Melvin F. Kanninen, MFK Consulting Services
Milton Levenson, Bechtel International (retired)
Allison Macfarlane, Massachusetts Institute of Technology
Richard A. Meserve, Carnegie Institution of Washington

Donald R. Olander, University of California, Berkeley
Theofanis G. Theofanous, University of California, Santa Barbara
George W. Ullrich, SAIC
Frank N. von Hippel, Princeton University

Although the reviewers listed above have provided many constructive comments and suggestions, they were not asked to endorse the report's conclusions or recommendations, nor did they see the final draft of the report before its release. The review of this report was overseen by Chris G. Whipple, ENVIRON International Corporation, and R. Stephen Berry, University of Chicago. Appointed by the National Research Council, they were responsible for making certain that an independent examination of this report was carried out in accordance with institutional procedures and that all review comments were carefully considered. Responsibility for the final content of this report rests entirely with the authoring committee and the institution.

CONTENTS

NOTE TO READERS

This report is based on a classified report that was developed at the request of the U.S. Congress with sponsorship from the Nuclear Regulatory Commission and the Department of Homeland Security. This report contains all of the findings and recommendations that appear in the classified report. Some have been slightly reworded and other sensitive information that might allow terrorists to exploit potential vulnerabilities has been redacted to protect national security. Nevertheless, the National Research Council and the authoring committee believe that this report provides an accurate summary of the classified report, including its findings and recommendations.

The authoring committee for this report examined the potential consequences of a large number of scenarios for attacking spent fuel storage facilities at commercial nuclear power plants. Some of these scenarios were developed by the Nuclear Regulatory Commission as part of its ongoing vulnerability analyses, whereas others were developed by the committee based upon the expertise of its members or suggestions from participants at the committee's open meetings. The committee focused its discussions about terrorist attacks on the concept of *maximum credible scenarios.* These are defined by the committee to be physically realistic classes of attacks that, if carried out successfully, would produce the most serious potential consequences within that class. In a practical sense they can be said to *bound* the consequences for a given type of attack. Such scenarios could in some cases be very difficult to carry out because they require a high level of skill and knowledge or luck on the part of the attackers. It was nevertheless useful to analyze these scenarios because they provide decision makers with a better understanding of the full range of potential consequences from terrorist attacks.

The committee uses the term *potential consequences* advisedly. It is important to recognize that a terrorist attack on a spent fuel storage facility would not necessarily result in the release of any radioactivity to the environment. The consequences of such an attack would depend not only on the nature of the attack itself, but also on the construction of the spent fuel storage facility; its location relative to surrounding features that might shield it from the attack; and the ability of the guards and operators at the facility to respond to the attack and/or mitigate its consequences. Facility-specific analyses are required to determine the potential vulnerability of a given facility to a given type of terrorist attack.

Congress asked the National Research Council for technical advice related to the vulnerability of spent fuel storage facilities to terrorist attacks. Congress, the Nuclear Regulatory Commission, and the Department of Homeland Security are responsible for translating this advice into policy actions. This will require the balancing of costs, risks, and benefits across the nation's industrial infrastructure. The committee was not asked to examine the potential vulnerabilities of other types of infrastructure to terrorist attacks or the consequences of such attacks. While such comparisons will likely be difficult, they will be essential for ensuring that the nation's limited resources are used judiciously in protecting its citizens from terrorist attacks.

SUMMARY FOR CONGRESS

The U.S. Congress asked the National Academies to provide independent scientific and technical advice on the safety and security of commercial spent nuclear fuel storage in the United States, specifically with respect to the following charges:

- Potential safety and security risks of spent nuclear fuel presently stored in cooling pools at commercial nuclear reactor sites.
- Safety and security advantages, if any, of dry cask storage versus wet pool storage at these reactor sites.
- Potential safety and security advantages, if any, of dry cask storage using various single-, dual-, and multi-purpose cask designs.
- The risks of terrorist attacks on these materials and the risk these materials might be used to construct a radiological dispersal device.

Congress requested that the National Academies produce a classified report that addresses these charges within 6 months and also provide an unclassified summary for unlimited public distribution. The first request was fulfilled in July 2004. This report fulfills the second request.

The highlights of the report are as follows:

(1) Spent fuel pools are necessary at all operating nuclear power plants to store recently discharged fuel.

(2) The committee judges that successful terrorist attacks on spent fuel pools, though difficult, are possible.

(3) If an attack leads to a propagating zirconium cladding fire, it could result in the release of large amounts of radioactive material.

(4) Additional analyses are needed to understand more fully the vulnerabilities and consequences of events that could lead to propagating zirconium cladding fires.

(5) It appears to be feasible to reduce the likelihood of a zirconium cladding fire by rearranging spent fuel assemblies in the pool and making provision for water-spray systems that would be able to cool the fuel, even if the pool or overlying building were severely damaged.

(6) Dry cask storage has inherent security advantages over spent fuel pool storage, but it can only be used to store older spent fuel.

(7) There are no large security differences among different storage-cask designs.

(8) It would be difficult for terrorists to steal enough spent fuel from storage facilities for use in significant radiological dispersal devices (dirty bombs).

The statement of task does not direct the committee to recommend whether the transfer of spent fuel from pool to dry cask storage should be accelerated. The committee judges, however, that further engineering analyses and cost-benefit studies would be needed before decisions on this and other mitigative measures are taken. The report contains detailed recommendations for improving the security of spent fuel storage regardless of how it is stored.

EXECUTIVE SUMMARY

In the Fiscal Year 2004 Energy and Water Development Conference Report, the U.S. Congress asked the National Academies to provide independent scientific and technical advice on the safety and security[1] of commercial spent nuclear fuel storage in the United States, specifically with respect to the following four charges:

(1) Potential safety and security risks of spent nuclear fuel presently stored in cooling pools at commercial reactor sites.
(2) Safety and security advantages, if any, of dry cask storage versus wet pool storage at these reactor sites.
(3) Potential safety and security advantages, if any, of dry cask storage using various single-, dual-, and multi-purpose cask designs.
(4) The risks of terrorist attacks on these materials and the risk these materials might be used to construct a radiological dispersal device.

Congress requested that the National Academies produce a classified report that addresses these charges within 6 months and also provide an unclassified summary for unlimited public distribution. The first request was fulfilled in July 2004. This report fulfills the second request.

Spent nuclear fuel is stored at commercial nuclear power plant sites in two configurations:

- In water-filled pools, referred to as *spent fuel pools*.
- In *dry casks* that are designed either for storage (single-purpose casks) or both storage and transportation (dual-purpose casks). There are two basic cask designs: bare-fuel casks and canister-based casks, which can be licensed for either single- or dual-purpose use, depending on their design.

Spent fuel pools are currently in use at all 65 sites with operating commercial nuclear power reactors, at 8 sites where commercial power reactors have been shut down, and at one site not associated with an operating or shutdown power reactor. Dry-cask storage facilities have been established at 28 operating, shutdown, or decommissioned power plants. The nuclear industry projects that up to three or four nuclear power plants will reach full capacity in their spent fuel pools each year for at least the next 17 years.

The congressional request for this study was prompted by conflicting public claims about the safety and security of commercial spent nuclear fuel storage at nuclear power plants. Some analysts have argued that the dense packing of spent fuel in cooling pools at nuclear power plants does not allow a sufficient safety margin in the event of a loss-of-pool-coolant event from an accident or terrorist attack. They assert that such events could result in the release of large quantities of radioactive material to the environment if the zirconium cladding of the spent fuel overheats and ignites. To reduce the potential for such fires, these

[1] In the context of this study, *safety* refers to measures that protect spent nuclear fuel storage facilities against failure, damage, human error, or other accidents that would disperse radioactivity in the environment. *Security* refers to measures to protect spent fuel storage facilities against sabotage, attacks, or theft.

analysts have suggested that spent fuel more than five years old be removed from the pool and stored in dry casks, and that the remaining younger fuel be reconfigured in the pool to allow more space for air cooling in the event of a loss-of-pool-coolant event.

The committee that was appointed to perform the present study examined the vulnerability of spent fuel stored in pools and dry casks to accidents and terrorist attacks. Any event that results in the breach of a spent fuel pool or a dry cask, whether accidental or intentional, has the potential to release radioactive material to the environment. The committee therefore focused its limited time on understanding two issues: (1) Under what circumstances could pools or casks be breached? And (2) what would be the radioactive releases from such breaches?

To address these questions, the committee performed a critical review of the security analyses that have been carried out by the Nuclear Regulatory Commission and its contractors, the Department of Homeland Security, industry, and other independent experts to determine if they are objective, complete, and credible. The committee was unable to examine several important issues related to these questions either because it was unable to obtain needed information from the Nuclear Regulatory Commission or because of time constraints. Details are provided in Chapters 1 and 2.

The committee's findings and recommendations from this analysis are provided below, organized by the four charges of the study task. The ordering of the charges has been rearranged to provide a more logical exposition of results.

CHARGE 4: RISKS OF TERRORIST ATTACKS ON THESE MATERIALS AND THE RISK THESE MATERIALS MIGHT BE USED TO CONSTRUCT A RADIOLOGICAL DISPERSAL DEVICE

The concept of *risk* as applied to terrorist attacks underpins the entire statement of task for this study. Therefore, the committee examined this final charge first to provide the basis for addressing the remainder of the task statement. The committee's examination of Charge 4 is provided in Chapter 2. On the basis of this examination, the committee offers the following findings and recommendations numbered according to the chapters in which they appear:

FINDING 2A: The probability of terrorist attacks on spent fuel storage cannot be assessed quantitatively or comparatively. Spent fuel storage facilities cannot be dismissed as targets for such attacks because it is not possible to predict the behavior and motivations of terrorists, and because of the attractiveness of spent fuel as a terrorist target given the well known public dread of radiation. Terrorists view nuclear power plant facilities as desirable targets because of the large inventories of radioactivity they contain. While it would be difficult to attack such facilities, the committee judges that attacks by knowledgeable terrorists with access to appropriate technical means are possible. It is important to recognize, however, that an attack that damages a power plant or its spent fuel storage facilities would not necessarily result in the release of *any* radioactivity to the environment. There are potential steps that can be taken to lower the potential consequences of such attacks.

FINDING 2B: The committee judges that the likelihood terrorists could steal enough spent fuel for use in a significant radiological dispersal device is small. Removal of a spent fuel assembly from the pool or dry cask would prove extremely difficult under almost any terrorist attack scenario. Attempts by a knowledgeable insider(s) to remove single rods and related debris from the pool might prove easier, but the amount of material that could be removed would be small. Moreover, superior materials could be stolen or purchased more easily from other sources. Even though the likelihood of spent fuel theft appears to be small, it is nevertheless important that the protection of these materials be maintained and improved as vulnerabilities are identified.

> **RECOMMENDATION: The Nuclear Regulatory Commission should review and upgrade, where necessary, its security requirements for protecting spent fuel rods not contained in fuel assemblies from theft by knowledgeable insiders, especially in facilities where individual fuel rods or portions of rods are being stored in pools.**

FINDING 2C: A number of security improvements at nuclear power plants have been instituted since the events of September 11, 2001. However, the Nuclear Regulatory Commission did not provide the committee with enough information to evaluate the effectiveness of these procedures for protecting stored spent fuel. Surveillance and other human-factors related security procedures are just as important as the physical barriers in preventing and mitigating terrorist attacks. Although the committee did learn about some of the changes that have been instituted since the September 11, 2001, attacks, it was not provided with enough information to evaluate the effectiveness of procedures now in place.

> **RECOMMENDATION: Although the committee did not specifically investigate the effectiveness and adequacy of improved surveillance and security measures for protecting stored spent fuel, an assessment of current measures should be performed by an independent[2] organization.**

CHARGE 1: POTENTIAL SAFETY AND SECURITY RISKS OF SPENT NUCLEAR FUEL STORED IN POOLS

The committee's examination of Charge 1 is provided in Chapter 3. On the basis of this examination, the committee offers the following findings and recommendations:

FINDING 3A: Pool storage is required at all operating commercial nuclear power plants to cool newly discharged spent fuel. Freshly discharged spent fuel generates too much decay heat to be passively air cooled. This fuel must be stored in a pool that has an active heat removal system (i.e., water pumps and heat exchangers) for at least one year before being moved to dry storage. Most dry storage systems are licensed to store fuel that has been out of the reactor for at least five years. Although spent fuel younger than five years could be stored in dry casks, the changes required for shielding and heat-removal

[2] That is, independent of the Nuclear Regulatory Commission and the nuclear industry.

could be substantial, especially for fuel that has been discharged for less than about three years.

FINDING 3B: The committee finds that, under some conditions, a terrorist attack that partially or completely drained a spent fuel pool could lead to a propagating zirconium cladding fire and the release of large quantities of radioactive materials to the environment. Details are provided in the committee's classified report.

FINDING 3C: It appears to be feasible to reduce the likelihood of a zirconium cladding fire following a loss-of-pool-coolant event using readily implemented measures. The following measures appear to have particular merit: Reconfiguring the spent fuel in the pools (i.e., redistribution of high decay-heat assemblies so that they are surrounded by low decay-heat assemblies) to more evenly distribute decay-heat loads and enhance radiative heat transfer; limiting the frequency of offloads of full reactor cores into spent fuel pools, requiring longer shutdowns of the reactor before any fuel is offloaded, and providing enhanced security when such offloads must be made; and development of a redundant and diverse response system to mitigate loss-of-pool-coolant events that would be capable of operation even if the pool or overlying building were severely damaged.

FINDING 3D: The potential vulnerabilities of spent fuel pools to terrorist attacks are plant-design specific. Therefore, specific vulnerabilities can be understood only by examining the characteristics of spent fuel storage at each plant. As described in Chapter 3, there are substantial differences in the designs of spent fuel pools that make them more or less vulnerable to certain types of terrorist attacks.

FINDING 3E: The Nuclear Regulatory Commission and independent analysts have made progress in understanding some vulnerabilities of spent fuel pools to certain terrorist attacks and the consequences of such attacks for releases of radioactivity to the environment. However, additional work on specific issues is needed urgently. The analyses carried out to date provide a general understanding of spent fuel behavior in a loss-of-pool-coolant event and the vulnerability of spent fuel pools to certain terrorist attacks that could cause such events to occur. The work to date, however, has not been sufficient to adequately understand the vulnerabilities and consequences of such events. Additional analyses are needed to fill in the knowledge gaps so that well-informed policy decisions can be made.

> **RECOMMENDATION: The Nuclear Regulatory Commission should undertake additional best-estimate analyses to more fully understand the vulnerabilities and consequences of loss-of-pool-coolant events that could lead to a zirconium cladding fire. Based on these analyses, the Commission should take appropriate actions to address any significant vulnerabilities that are identified.** The committee provides details on additional analyses that should be carried out in its classified report. Cost-benefit considerations will be an important part of such decisions.

> **RECOMMENDATION: While the work described in the previous recommendation under Finding 3E, above, is being carried out, the Nuclear Regulatory Commission should ensure that power plant operators take prompt and effective measures to reduce the consequences of loss-of-pool-coolant**

events in spent fuel pools that could result in propagating zirconium cladding fires. The committee judges that there are at least two such measures that should be implemented promptly:

- Reconfiguring of fuel in the pools so that high decay-heat fuel assemblies are surrounded by low decay-heat assemblies. This will more evenly distribute decay-heat loads, thus enhancing radiative heat transfer in the event of a loss of pool coolant.
- Provision for water-spray systems that would be able to cool the fuel even if the pool or overlying building were severely damaged.

Reconfiguring of fuel in the pool would be a prudent measure that could probably be implemented at all plants at little cost, time, or exposure of workers to radiation. The second measure would probably be more expensive to implement and may not be needed at all plants, particularly plants in which spent fuel pools are located below grade or are protected from external line-of-sight attacks by exterior walls and other structures.

The committee anticipates that the costs and benefits of options for implementing the second measure would be examined to help decide what requirements would be imposed. Further, the committee does not presume to anticipate the best design of such a system—whether it should be installed on the walls of a pool or deployed from a location where it is unlikely to be compromised by the same attack—but simply notes the demanding requirements such a system must meet.

CHARGE 3: POTENTIAL SAFETY AND SECURITY ADVANTAGES, IF ANY, OF DIFFERENT DRY CASK STORAGE DESIGNS

The third charge to the committee focuses exclusively on the safety and security of dry casks. The committee addressed this charge first in Chapter 4 to provide the basis for the comparative analysis between dry casks and pools as called for in Charge 2.

FINDING 4A: Although there are differences in the robustness of different dry cask designs (e.g., bare-fuel versus canister-based), the differences are not large when measured by the absolute magnitudes of radionuclide releases in the event of a breach. All storage cask designs are vulnerable to some types of terrorist attacks, but the quantity of radioactive material releases predicted from such attacks is relatively small. These releases are not easily dispersed in the environment.

FINDING 4B: Additional steps can be taken to make dry casks less vulnerable to potential terrorist attacks. Although the vulnerabilities of current cask designs are already small, additional, relatively simple steps can be taken to reduce them as discussed in Chapter 4.

RECOMMENDATION: The Nuclear Regulatory Commission should consider using the results of the vulnerability analyses for possible upgrades of requirements in 10 CFR 72 for dry casks, specifically to improve their resistance to terrorist attacks. The committee was told by

Nuclear Regulatory Commission staff that such a step is already under consideration.

CHARGE 2: SAFETY AND SECURITY ADVANTAGES, IF ANY, OF DRY CASK STORAGE VERSUS WET POOL STORAGE

In Chapter 4, the committee offers the following findings and recommendations with respect to the comparative component of Charge 2:

FINDING 4C: Dry cask storage does not eliminate the need for pool storage at operating commercial reactors. Under present U.S. practices, dry cask storage can only be used to store fuel that has been out of the reactor long enough (generally greater than five years under current practices) to be passively air cooled.

FINDING 4D: Dry cask storage for older, cooler spent fuel has two inherent advantages over pool storage: (1) It is a passive system that relies on natural air circulation for cooling; and (2) it divides the inventory of that spent fuel among a large number of discrete, robust containers. These factors make it more difficult to attack a large amount of spent fuel at one time and also reduce the consequences of such attacks. The robust construction of these casks prevents large-scale releases of radioactivity in all of the attack scenarios examined by the committee in its classified report.

FINDING 4E: Depending on the outcome of plant-specific vulnerability analyses described in the committee's classified report, the Nuclear Regulatory Commission might determine that earlier movements of spent fuel from pools into dry cask storage would be prudent to reduce the potential consequences of terrorist attacks on pools at some commercial nuclear plants. The statement of task directs the committee to examine the risks of spent fuel storage options and alternatives for decision makers, not to recommend whether any spent fuel should be transferred from pool storage to cask storage. In fact, there may be some commercial plants that, because of pool designs or fuel loadings, may require some removal of spent fuel from their pools. If there is a need to remove spent fuel from the pools it should become clearer once the vulnerability and consequence analyses described in the classified report are completed. The committee expects that cost-benefit considerations would be a part of these analyses.

IMPLEMENTATION ISSUES

Implementation of the recommendations in Chapters 2-4 will require action and cooperation by a large number of parties. The final chapter of the report provides a brief discussion of two implementation issues that the committee believes are of special interest to Congress: *Timing Issues*: Ensuring that high-quality, expert analyses are completed in a timely manner; and *Communications Issues*: Ensuring that the results of the analyses are communicated to relevant parties so that appropriate and timely mitigating actions can be taken. This discussion leads to the following finding and recommendation.

FINDING 5A: Security restrictions on sharing of information and analyses are hindering progress in addressing potential vulnerabilities of spent fuel storage to

terrorist attacks. Current classification and security practices appear to discourage information sharing between the Nuclear Regulatory Commission and industry. They impede the review and feedback processes that can enhance the technical soundness of the analyses being carried out; they make it difficult to build support within the industry for potential mitigative measures; and they may undermine the confidence that the industry, expert panels such as this one, and the public place in the adequacy of such measures.

> **RECOMMENDATION: The Nuclear Regulatory Commission should improve the sharing of pertinent information on vulnerability and consequence analyses of spent fuel storage with nuclear power plant operators and dry cask storage system vendors on a timely basis.**

The committee also believes that the public is an important audience for the work being carried out to assess and mitigate vulnerabilities of spent fuel storage facilities. While it would be inappropriate to share all information publicly, more constructive interaction with the public and independent analysts could improve the work being carried out and also increase public confidence in Nuclear Regulatory Commission and industry decisions and actions to reduce the vulnerability of spent fuel storage to terrorist threats.

1
INTRODUCTION AND BACKGROUND

In the Fiscal Year 2004 Energy and Water Development Conference Report, the U.S. Congress asked the National Academies to provide independent scientific and technical advice on the safety and security[1] of commercial spent nuclear fuel storage in the United States (see Box 1.1). The Nuclear Regulatory Commission and the Department of Homeland Security jointly sponsored this study, as directed by Congress.

Awareness and concerns about the threat of high-impact terrorism have become acute and pervasive since the attacks on September 11, 2001. The information gathered by the committee during this study led it to conclude that there were indeed credible concerns about the safety and security of spent nuclear fuel storage in the current threat environment. From the outset the committee believed that safety and security issues must be addressed quickly to determine whether additional measures are needed to prevent or mitigate attacks that could cause grave harm to people and cause widespread fear, disruption, and economic loss. The information gathered during this study reinforced that view. Any concern related to nuclear power plants[2] has added stakes: Many people fear radiation more than they fear exposure to other physical insults. This amplifies the concern over a potential terrorist attack involving radioactive materials beyond the physical injuries it might cause, and beyond the economic costs of the cleanup.

1.1 CONTEXT FOR THIS STUDY

The congressional request for this study was prompted by conflicting public claims about the safety and security of commercial spent nuclear fuel storage at nuclear power plants. Some have argued that the dense packing used for storing spent fuel in cooling pools at nearly every nuclear power plant does not provide a sufficient safety margin in the event of a pool breach and consequent water loss from an accident or terrorist attack.[3] In such cases, the potential exists for the fuel most recently discharged from a reactor to heat up sufficiently for its zirconium cladding to ignite, possibly resulting in the release of large amounts of radioactivity to the environment (Alvarez et al., 2003a). The Nuclear Regulatory Commission's own analyses have suggested that such zirconium cladding fires and releases of radioactivity are possible (e.g., USNRC, 2001a).

To reduce the potential for such an event, Alvarez et al. (2003a) suggested that spent fuel more than five years old be removed from the pool and stored in dry casks, and

---

[1] In the context of this study, *safety* refers to measures that protect spent nuclear fuel storage facilities against failure, damage, human error, or other accidents that would disperse radioactivity in the environment. *Security* refers to measures to protect spent fuel storage facilities against sabotage, attacks, or theft.
[2] Safety and security of reactors at nuclear power plants are outside of the committee's statement of task and have been addressed only where they could not be separated from spent fuel storage. The distinctions between spent fuel storage and operating nuclear power reactors are sometimes blurred in public discussions of nuclear and radiological concerns.
[3] The committee refers to such occurrences as *loss-of-pool-coolant events* in this report.

BOX 1.1 STATEMENT OF TASK

The issues to be addressed by this study are specified in the Energy and Water Development Conference Report and are as follows:

(1) Potential safety and security risks of spent nuclear fuel presently stored in cooling pools at commercial reactor sites (see Chapter 3).
(2) Safety and security advantages, if any, of dry cask storage versus wet pool storage at these reactor sites (see Chapter 4).
(3) Potential safety and security advantages, if any, of dry cask storage using various single-, dual-, and multi-purpose cask designs (see Chapter 4).
(4) In light of the September 11, 2001, terrorist attacks, this study will explicitly consider the risks of terrorist attacks on these materials and the risk these materials might be used to construct a radiological dispersal device (see Chapter 2).

that the remaining younger fuel be rearranged in the pool to allow more space for cooling (see also Marsh and Stanford, 2001; Thompson, 2003). The Nuclear Regulatory Commission staff, the nuclear industry, and some others have argued that densely packed pool storage can be carried out both safely and securely (USNRC, 2003a).

Policy actions to improve the safety and security of spent fuel storage could have significant national consequences. Nuclear power plants generate approximately 20 percent of the electricity produced in the United States. The issue of its future availability and use is critical to our nation's present and future energy security. The safety and security of spent fuel storage is an important aspect of the acceptability of nuclear power. Decisions that affect such a large portion of our nation's electricity supply must be considered carefully, wisely, and with a balanced view.

1.2 STRATEGY TO ADDRESS THE STUDY CHARGES

Congress directed the National Academies to produce a classified report that addresses the statement of task shown in Box 1.1 within 6 months and an unclassified summary for unlimited public dissemination within 12 months. This report, which has undergone a security review by the Nuclear Regulatory Commission and found to contain no classified national security or safeguards information, fulfills the second request.[4]

The National Research Council of the National Academies appointed a committee of 15 experts to carry out this study. Biographical sketches of the committee members are provided in Appendix B. The committee met six times from February to June 2004 to gather information and complete its classified report. The committee met again in August, October, and November 2004 and in January 2005 to develop this public report.

Details on the information-gathering sessions and speakers are provided in Appendix A. Most of the information-gathering sessions were not open to the public because they involved presentations and discussions of classified information. The committee recognized, however, that important contributions to this study could be made by industry representatives, independent analysts, and the public, so it scheduled open, unclassified

[4] The classified report was briefed to the agencies and Congress on July 15, 2004.

sessions at three of its meetings to obtain comments from interested organizations and individuals. Public comments at these meetings were encouraged and considered.

Subgroups of the committee visited several nuclear power plants to learn first-hand how spent fuel is being managed in wet and dry storage: the Dresden and Braidwood Nuclear Generating Stations in Illinois, which are owned and operated by Exelon Nuclear Corp.; the Indian Point Nuclear Generating Station in New York, which is owned and operated by ENTERGY Corp.; and the Palo Verde Nuclear Generating Station in Arizona, which is operated by Arizona Public Service Corp. A subgroup of committee members also traveled to Germany to visit spent fuel storage installations at Ahaus and Lingen and to talk with experts about the safety and security of German spent fuel storage. The German government has been concerned about security for a long time, and the German nuclear industry has made adjustments to spent fuel storage designs and operations that reduce their vulnerability to accidents and terrorist attacks. A summary of the trip to Germany is provided in Appendix C.

The statement of task for this study directed the committee to examine both the safety and the security of spent fuel storage. It is important to recognize that these are two sides of the same coin in the sense that any event that results in the breach of a spent fuel pool or a dry cask, whether accidental or intentional, has the potential to release radioactive material to the environment. The committee therefore focused its limited time on understanding two issues: (1) Under what circumstances could pools or casks be breached? And (2) what would be the radioactive releases from such breaches?

The initiating events that could lead to the *accidental* breach of a spent fuel pool are well known: A large seismic event or the accidental drop of a cask on the pool wall that could lead to the loss of pool coolant. The condition that could lead to an accidental breach of a dry storage cask is similarly well known: an accidental drop of the cask during handling operations. Current Nuclear Regulatory Commission regulations are designed to prevent such accidental conditions by imposing requirements on the design and operation of spent fuel storage facilities. These regulations have been in place for decades and have so far been effective in preventing accidental releases of radioactive materials from these facilities into the environment.

The initiating events that could lead to the *intentional* breach of a spent fuel pool or dry storage cask are not as well understood. The Nuclear Regulatory Commission has had long-standing requirements in place to deal with radiological sabotage (included in the "design basis threat"; see Chapter 2), but the September 11, 2001, terrorist attacks provided a graphic demonstration of a much broader array of potential threats. As described in the following chapters, the Nuclear Regulatory Commission is currently sponsoring studies to better understand the potential consequences of such terrorist attacks on spent fuel storage facilities.

Early on in this study, the committee made a judgment that it should focus most of its attention concerning such initiating events on the security aspects of its task statement. Many of the phenomena that follow an initiating event (e.g., loss of pool coolant or cask breach) would be the same whether it arose from an accident or terrorist attack, as noted previously. While the mitigation strategies for such events might be similar, they would require different kinds of preparation.

Given the relatively short time frame for this study, the committee focused its efforts on performing a critical review of the security analyses that have been carried out by the Nuclear Regulatory Commission and its contractors, the Department of Homeland Security, industry (i.e., EPRI, formerly named the Electric Power Research Institute; ENTERGY Corp.; and dry cask vendors), and other independent experts to determine if they are objective, complete, and credible. The committee could only perform limited independent safety and security analyses based on the information it gathered.

The committee made many requests for information from the Nuclear Regulatory Commission, its Sandia National Laboratories contractor, and other organizations and individuals, often with little advance notice. For the most part, all parties responded well to these requests. The committee was able to access experts who could answer its technical questions and was pleased with the cooperation and information it received during its visits to spent fuel storage facilities. This cooperation was essential in enabling the committee to complete its task within the requested six-month timeframe.

The committee was forced to circumscribe some aspects of its examinations, however, due to time and/or information constraints. In particular, the committee did not pursue in-depth examinations of the following topics:

- Human factors issues involved in responding to terrorist attacks on spent fuel storage. These include surveillance activities to identify potential threats (both inside and outside the plant); the response of security forces; and the preparation of plant personnel to deploy mitigative measures in the event of an attack.
- The behavior of radioactive material after it enters the environment from a spent fuel pool or dry cask. The committee assumed that any large release of radioactivity from a spent fuel storage facility would be problematic even in the absence of knowledge of how it would disperse in the environment. The committee instead focused its efforts on understanding how much radioactive material would be released, if any, in the case of an attack.
- The economic consequences of potential terrorist attacks, except insofar as noting the possible magnitude of cleanup costs after a catastrophic release of radioactivity.
- The costs of potential measures to mitigate spent fuel storage vulnerabilities. The committee understands that the Nuclear Regulatory Commission would include cost-benefit considerations in decisions to impose any new requirements on industry for such measures.

The committee also did not examine the potential vulnerability of commercial spent fuel while being transported. That topic is not only outside of the committee's task, but there is another National Academies study currently underway to examine transportation issues.[5]

Because most of the studies on spent fuel storage vulnerabilities undertaken for the Nuclear Regulatory Commission are still in progress, the committee was not able to review completed technical documents. Instead, the committee had to rely on presentations by and discussions with technical experts. The committee does not believe that these difficulties prevented it from developing sound findings and recommendations from the information it

---

[5] Committee on Transportation of Radioactive Waste. See *http://national-academies.org/transportofradwaste*. That committee's final report is now planned for completion in the late summer of 2005.

did receive. The committee was able to draw upon other information sources both domestic and foreign,[6] including the experience and expertise of its members, to fill some of the information gaps.

1.3 REPORT ROADMAP

The sections that follow in this chapter provide background on storage of spent nuclear fuel, which may be helpful to non-experts in understanding the issues discussed in the following chapters. The other chapters are organized to explicitly address the four charges of the committee's statement of task:

- Chapter 2 addresses the last charge to the committee to "explicitly consider the risks of terrorist attacks on these materials and the risk these materials might be used to construct a radiological dispersal device."
- Chapter 3 addresses the first charge to the committee to examine the "potential safety and security risks of spent nuclear fuel presently stored in cooling pools at commercial reactor sites."
- Chapter 4 addresses the second and third charges to examine the "safety and security advantages, if any, of dry cask storage versus wet pool storage at these reactor sites" and the "potential safety and security advantages, if any, of dry cask storage using various single-, dual-, and multi-purpose cask designs."
- Chapter 5 concerns implementation of the recommendations in this report, specifically concerning timing and communication issues.

The appendixes provide supporting information, including a glossary and acronym list, descriptions of the committee's meetings, and biographical sketches of the committee members.

1.4 BACKGROUND ON SPENT NUCLEAR FUEL AND ITS STORAGE

This section is provided for readers who are not familiar with the technical features of spent nuclear fuel and its storage. Other readers should skip directly to Chapter 2.

Spent nuclear fuel is fuel that has been irradiated or "burned" in the core of a nuclear reactor. In power reactors, the energy released from fission reactions in the nuclear fuel heats water[7] to produce steam that drives turbines to generate electricity. Spent nuclear fuel from non-commercial reactors (such as research reactors, naval propulsion reactors, and plutonium production reactors) is not considered in this study.

1.4.1 Nuclear Fuel

Almost all commercial reactor fuel in the United States is in the form of solid, cylindrical pellets of uranium dioxide. The pellets are about 0.4 to 0.65 inch (1.0 to 1.65 centimeters) in length and about 0.3 to 0.5 inch (0.8 to 1.25 centimeters) in diameter. The

---

[6] For example, the aforementioned visits to Lingen and Ahaus, in Germany.

[7] A different coolant can be used, but all power reactors now operating in the United States are water cooled.

pellets are loaded into tubes, called *fuel cladding*, made of a zirconium metal alloy, called zircaloy. A loaded tube, which is typically 11.5 to 14.75 feet (3.5 to 4.5 meters) in length, is called a *fuel rod* (also referred to as a *fuel pin* or *fuel element*). Fuel rods are bundled together, with a 0.12 to 0.18 inch (0.3 to 0.45 centimeter) space left between each for coolant to flow, to form a square fuel assembly (see FIGURE 1.1) measuring about 6 to 9 inches (15 to 23 centimeters) on a side.

Typical fuel assemblies for boiling water nuclear reactors (BWRs) hold 49 to 63 fuel rods, and fuel assemblies for pressurized water nuclear reactors (PWRs) hold 164 to 264 fuel rods.[8] Depending on reactor design, typically between 190 and 750 assemblies, each weighing from 275 to 685 kg (600 to 1500 pounds), make up a power reactor core. New fuel assemblies (i.e., those that have not been irradiated in a reactor) do not require special cooling or radiation shielding; they can be moved with a crane in open air. Once in the reactor, however, the fuel undergoes nuclear fission and begins to generate the radioactive fission products and activation products that require shielding and cooling.

The uranium oxide fuel essentially is composed of two isotopes of uranium: Initially, about 3-5 percent[9] by weight is fissile uranium (uranium-235), which is the component that sustains the fission chain reaction; and about 95-97 percent is uranium-238, which can capture a neutron to produce fissile plutonium and other radioactive heavy isotopes (actinides). Each fission event, whether in uranium or plutonium, releases energy and neutrons as the fissioning nucleus splits into two (and infrequently three) radioactive fragments, called fission products.

When the fissile material has been consumed to a level where it is no longer economically viable (typically 4.5 to 6 years of operation for current fuel designs), the fuel is considered *spent* and is removed from the reactor core. Spent fuel assemblies are highly radioactive. The decay of radioactive fission products and other constituents generates heat (called *decay heat*) and penetrating (gamma and neutron) radiation. Therefore cooling, shielding, and remote handling are required for spent nuclear fuel.

The amount of heat and radiation generated by a spent fuel assembly after its removal from a reactor depends on the number of fissions that have occurred in the fuel, called the *burn-up*, and the time that has elapsed since the fuel was removed from the reactor. The rate of decay-heat generation by spent reactor fuel and how it will change with time after the fuel is removed from the reactor can be calculated. The results of an example calculation are shown in FIGURE 1.2.

At discharge from the reactor, a spent fuel assembly generates on the order of tens of kilowatts of heat. Decay-heat production diminishes as very short-lived radionuclides decay away, dropping heat generation by a factor of 100 during the first year; dropping by another factor of 5 between year one and year five; and dropping about 40 percent between year five and year ten (see FIGURE 1.2). Within a year of discharge from the reactor, decay-heat production in spent nuclear fuel is dominated by four radionuclides: Ruthenium-106 (with a 372.6-day half-life), cerium-144 (284.4-day half-life), cesium-137 (30.2-year half-life),

---

[8] Technical specifications for the fuel assemblies are taken from the American National Standard document for pool storage of spent nuclear fuel (American Nuclear Society, 1988).

[9] With only a few exceptions, commercial nuclear power reactors in the United States have been fueled with low-enriched uranium, that is, less than 20 percent of the uranium is uranium-235. Uranium found in nature has about 0.71 percent uranium-235 by weight.

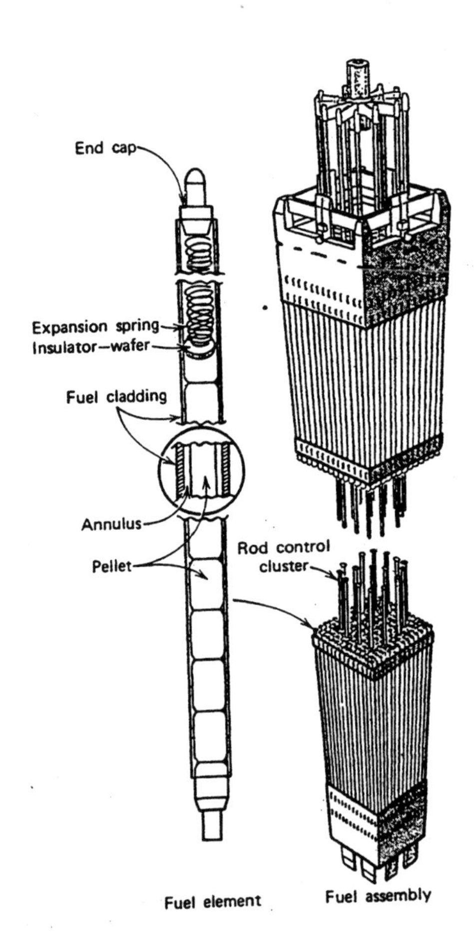

FIGURE 1.1 Fuel rods, also called fuel pins or elements, are bundled together into fuel assemblies as shown here. This fuel assembly is for a PWR reactor. SOURCE: Duderstadt and Hamilton (1976; Figure 3-7).

and cesium-134 (2.1-year half-life) and their short-lived decay products contribute nearly 90 percent of the decay heat from a spent fuel assembly.

Longer-lived radionuclides persist in the spent fuel even as the decay heat drops further. Cesium-137 decays to barium-137, emitting a beta particle and a high-energy gamma ray. The cesium-137 half-life of 30.2 years is sufficiently long to ensure that this radionuclide will persist during storage. It and other materials present in the fuel will form small particles, called *aerosols*, in a zirconium cladding fire.

Shorter-lived radionuclides decay away rapidly after removal of the spent fuel from the reactor. One of these is iodine-131, which is of particular concern in reactor core accidents because it can be taken up in large quantities by the human thyroid. This radionuclide has a half-life of about 8 days and typically persists in significant quantities in spent fuel only on the order of a few months.

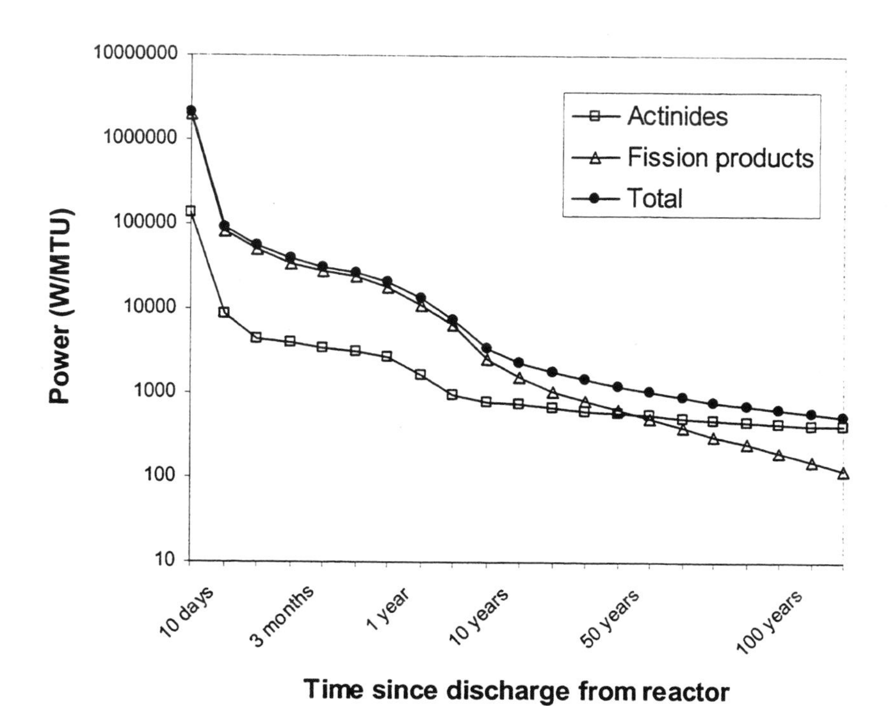

FIGURE 1.2 Decay-heat power for spent fuel (measured in watts per metric ton of uranium) plotted on a logarithmic scale as a function of time after reactor discharge. Note that the horizontal axis is a data series, not a scale. SOURCE: Based on data from USNRC (1984).

1.4.2 Storage of Spent Nuclear Fuel

Storage technologies for spent nuclear fuel have three primary objectives:

- Cool the fuel to prevent heat-up to high temperatures from radioactive decay.
- Shield workers and the public from the radiation emitted by radioactive decay in the spent fuel and provide a barrier for any releases of radioactivity.
- Prevent criticality accidents (uncontrolled fission chain reactions).

After the fuel assemblies are unloaded from the reactor they are stored in water pools, called *spent fuel pools*. The water in the pools provides radiation shielding and cooling and captures all but noble gas radionuclides in case of fuel rod leaks.[10] The geometry of the fuel and neutron absorbers (such as boron, hafnium, and cadmium) within the racks that hold the spent fuel or in the cooling water help prevent criticality events.[11] The water in the pool is circulated through heat exchangers for cooling and ion exchange filters to capture any radionuclides and other contaminants that get into the water. Makeup water is also added to the pool to replace pool water lost to evaporation. The operation of the pumps and heat exchangers is especially important during and immediately after reactor

[10] If the cladding in the fuel rods is breached some radioactive materials will be released into the pool.
[11] See the Glossary (Appendix E) for a definition of criticality. Most of the fuel's capacity for sustaining criticality is expended in the reactor as the uranium and plutonium are fissioned.

refueling operations, because this is when larger quantities of higher heat-generating spent fuel are placed into the pool.

Current U.S. regulations require that spent fuel be stored in the power plant's fuel pool for at least one year after its discharge from the reactor before being moved to dry storage. After that time the spent fuel can be moved, but only with active cooling. Active cooling is generally necessary for about three years after the spent fuel is removed from the reactor core (USNRC, 2003b).

When a spent fuel pool is filled to capacity, older fuel, which has lower decay-heat, is moved to other pools or placed into dry casks. Heat generated in the loaded dry casks is removed by air convection and thermal radiation. The cask provides shielding of penetrating radiation and confinement of the radionuclides in the spent fuel. As with pool storage, criticality control is accomplished by placing the fuel in a fixed geometry and separating individual fuel assemblies with neutron absorbers. Standard industry practice is to place in dry storage only spent fuel that has cooled for five years or more after discharge from the reactor.[12] Most spent fuel in wet or dry storage is located at nuclear power plant sites (i.e., on-site storage).

There are significant differences in the design and construction of wet and dry storage installations at commercial nuclear power plants. The characteristics depend on the type of the nuclear power plant, the age of the spent fuel storage installation, or the type of dry casks used. The design and features of spent fuel pools and dry storage facilities are discussed in Chapters 3 and 4, respectively.

1.4.3 Spent Fuel Inventories

As of 2003, approximately 50,000 MTU (metric tons of uranium) of spent fuel have been generated over the past four decades in the United States. A typical nuclear power plant generates about 20 MTU per year. The entire U.S. nuclear industry generates about 2000 MTU per year.

Of the approximately 50,000 MTU of commercial spent fuel in the United States, 43,600 MTU are currently stored in pools and 6200 MTU are in dry storage. Pool storage exists at all 65 sites with operating commercial nuclear power reactors[13] and at 8 sites where commercial power reactors are no longer operating (i.e., they have been shut down or decommissioned) (FIGURE 1.3). Additionally, there is an away-from-reactor spent fuel pool operating at the G.E. Morris Facility in Illinois (see Appendix D).

Of the spent fuel in dry storage, 4500 MTU are in storage at 22 sites with operating commercial nuclear power reactors, and 1700 MTU are in storage at 6 sites where the commercial reactors are no longer operating. An additional dry-storage facility is operated by the federal government at the Idaho National Laboratory. It stores most of the damaged fuel from the Three Mile Island Unit 2 reactor accident.

---

[12] Fuel aged as little as three years could be stored in passively cooled casks, but fewer assemblies could be accommodated in each cask because of the higher heat load.

[13] There are 103 operating commercial nuclear power reactors in the United States. Many sites have more than one operating reactor.

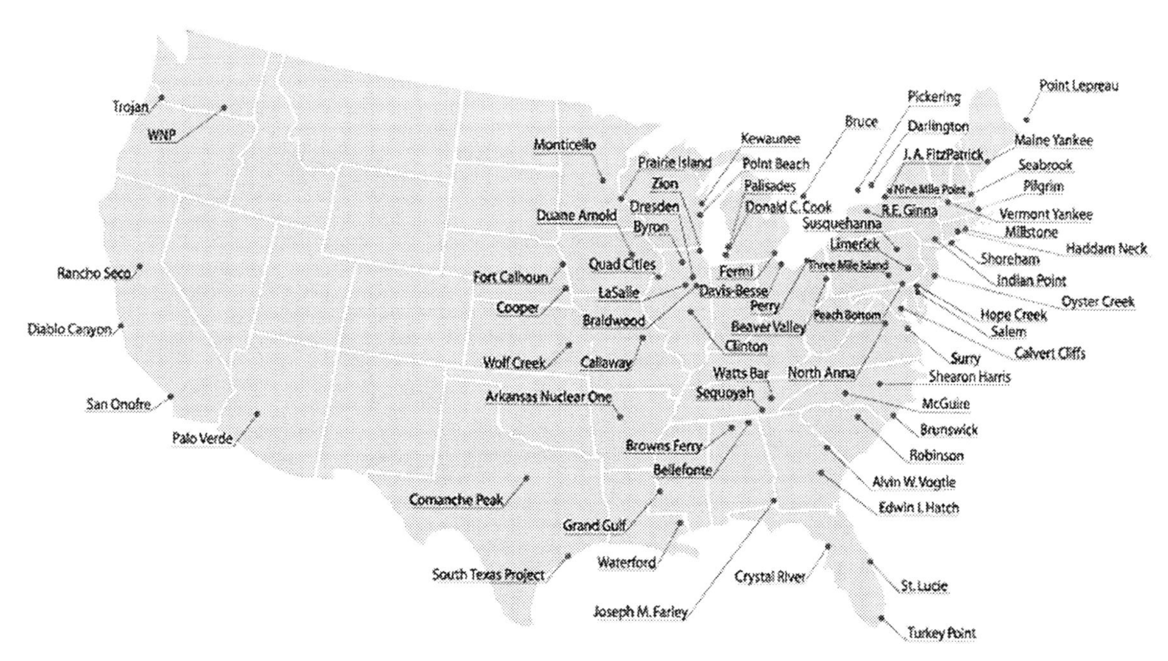

FIGURE 1.3 Locations of spent fuel storage facilities in the United States.

TABLE 1.1 provides a listing of the 30 operating Independent Spent Fuel Storage Installations (ISFSIs[14]) in the United States. These ISFSIs include the dry storage facilities at operating and shutdown commercial power reactor sites as well as the storage facilities at the Morris and Idaho sites, as described above. The committee did not examine the Morris and Idaho facilities as part of this study. At-reactor pool storage is not considered to be an ISFSI because it operates under the power reactor license.

1.4.4 History of Spent Fuel Storage

Spent fuel pools at commercial nuclear power plants were not designed to accommodate all the fuel used during the operating lifetime of the reactors they service. Most commercial power plants were designed with small pools under the assumption that fuel would be cooled for a short period of time after discharge from the reactor and then be sent offsite for recycling (i.e., reprocessing).[15] A commercial reprocessing industry never developed, however, for the reasons discussed in Appendix D. Newer power plants were designed with larger pool storage capacities. Even plants with larger-capacity pools will run out of pool space if they operate beyond their initial 40-year licenses. In 2000, the nuclear power industry projected that roughly three or four plants per year would run out of needed storage space in their pools without additional interim storage capacity (see FIGURE 1.4).

Another development that logically could reduce the demand for storage of spent nuclear fuel at the sites of power plants is the availability of a geologic repository for

---

[14] An ISFSI is a facility for storing spent fuel in wet pools or dry casks and is defined in Title 10, Part 72 of the Code of Federal Regulations.
[15] Residual uranium-235 and plutonium in the spent fuel would be recovered for the manufacture of new fuel. The waste products in the fuel, principally the fission products, would be immobilized in solid matrices and stored for eventual disposal.

TABLE 1.1: Operating ISFSIs in the United States as of July 2004

Name	Location
Palo Verde	Arizona
Arkansas Nuclear One	Arkansas
Rancho Seco	California
San Onofre	California
Diablo Canyon	California
Fort St. Vrain [1]	Colorado
Edwin L. Hatch	Georgia
DOE-INL [2]	Idaho
G.E. Morris [3]	Illinois
Dresden	Illinois
Duane Arnold	Iowa
Maine Yankee	Maine
Calvert Cliffs	Maryland
Big Rock Point	Michigan
Palisades	Michigan
Prairie Island	Minnesota
Yankee Rowe	Massachusetts
Oyster Creek	New Jersey
J.A. FitzPatrick	New York
McGuire	North Carolina
Davis-Besse	Ohio
Trojan	Oregon
Susquehanna	Pennsylvania
Peach Bottom	Pennsylvania
Robinson	South Carolina
Oconee	South Carolina
North Anna	Virginia
Surry	Virginia
Columbia Gen. Station	Washington
Point Beach	Wisconsin

NOTES:
[1]The Fort St. Vrain ISFSI stores fuel from a commercial gas-cooled reactor. The facility is operated by the Department of Energy.
[2]The DOE-INL facility stores fuel from the Three-Mile Island Unit 2 reactor. The facility is operated by the Department of Energy.
[3]The G.E. Morris ISFSI is a wet storage facility.

SOURCES: Data from the USNRC (2004).

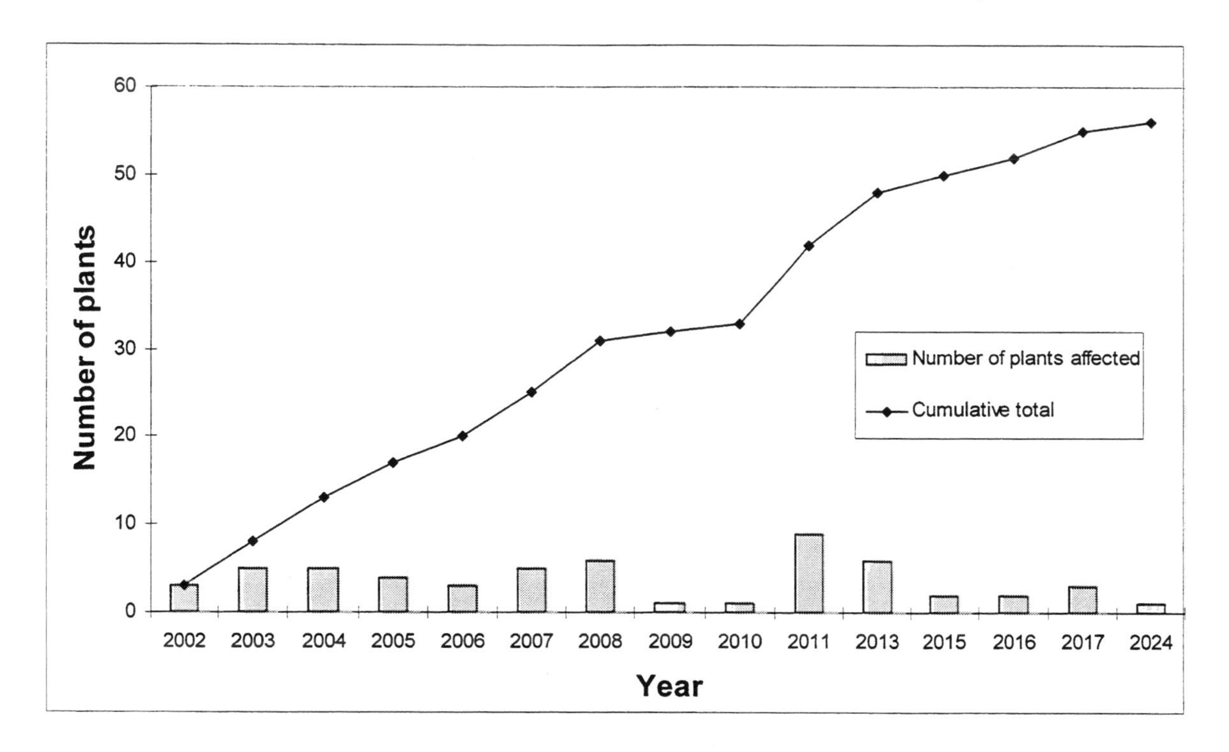

FIGURE 1.4 Projection of the number of commercial nuclear power plants that will run out of needed space in their spent fuel pools in coming years if they do not add interim storage. These data, looking only at plants that did not already use dry cask storage, were provided to the Nuclear Regulatory Commission in 2000. SOURCE: USNRC (2001b).

disposal of spent nuclear fuel. But a nuclear waste repository is not expected to be in operation until at least 2010, and even then it will take several decades for all of the spent fuel to be shipped for disposal. Thus, onsite storage of spent fuel is likely to continue for at least several decades.

Power plant operators have made two changes in spent fuel storage procedures to increase the capacity of onsite storage. First, starting in the late 1970s, plant operators began to install high-density racks that enable more spent fuel to be stored in the pools. This has increased storage capacities in some pools by up to about a factor of five (USNRC, 2003b). Second, as noted above, many plant operators have moved older spent fuel from the pools into dry cask storage systems (see Chapter 4) or into other pools when available to make room for freshly discharged spent fuel and to maintain the capacity for a full-core offload.[16]

The original spent fuel racks, sometimes called "open racks," were designed to store spent fuel in an open array, with open vertical and lateral channels between the fuel assemblies to promote water circulation. The high-density storage racks eliminated many of the channels so that the fuel assemblies could be packed closer together (FIGURE 1.5). This configuration does not allow as much water (or air circulation in loss-of-pool-coolant events) through the spent fuel assemblies as the original open-rack design.

---

[16] Although not required by regulation, it is standard practice in the nuclear industry to maintain enough open space in the spent fuel pool to hold the entire core of the nuclear reactor. This provides an additional margin of safety should the fuel have to be removed from the reactor core in an emergency or for maintenance purposes.

Several nuclear utilities have already submitted license applications to the Nuclear Regulatory Commission to build 16 new ISFSIs. Among the potential new ISFSIs, a consortium of utilities has submitted a license for a private fuel storage facility (PFS) in Utah for interim dry storage of up to 40,000 metric tons of spent fuel.

Most or all pools store some spent fuel that has aged more than five years after discharge from the reactor, and so could be transferred to dry-cask storage. The amount that could be transferred depends on plant-specific information such as pool size and configuration, operating history of the reactor, the enrichment and burn-up level in the fuel, and availability of an ISFSI.

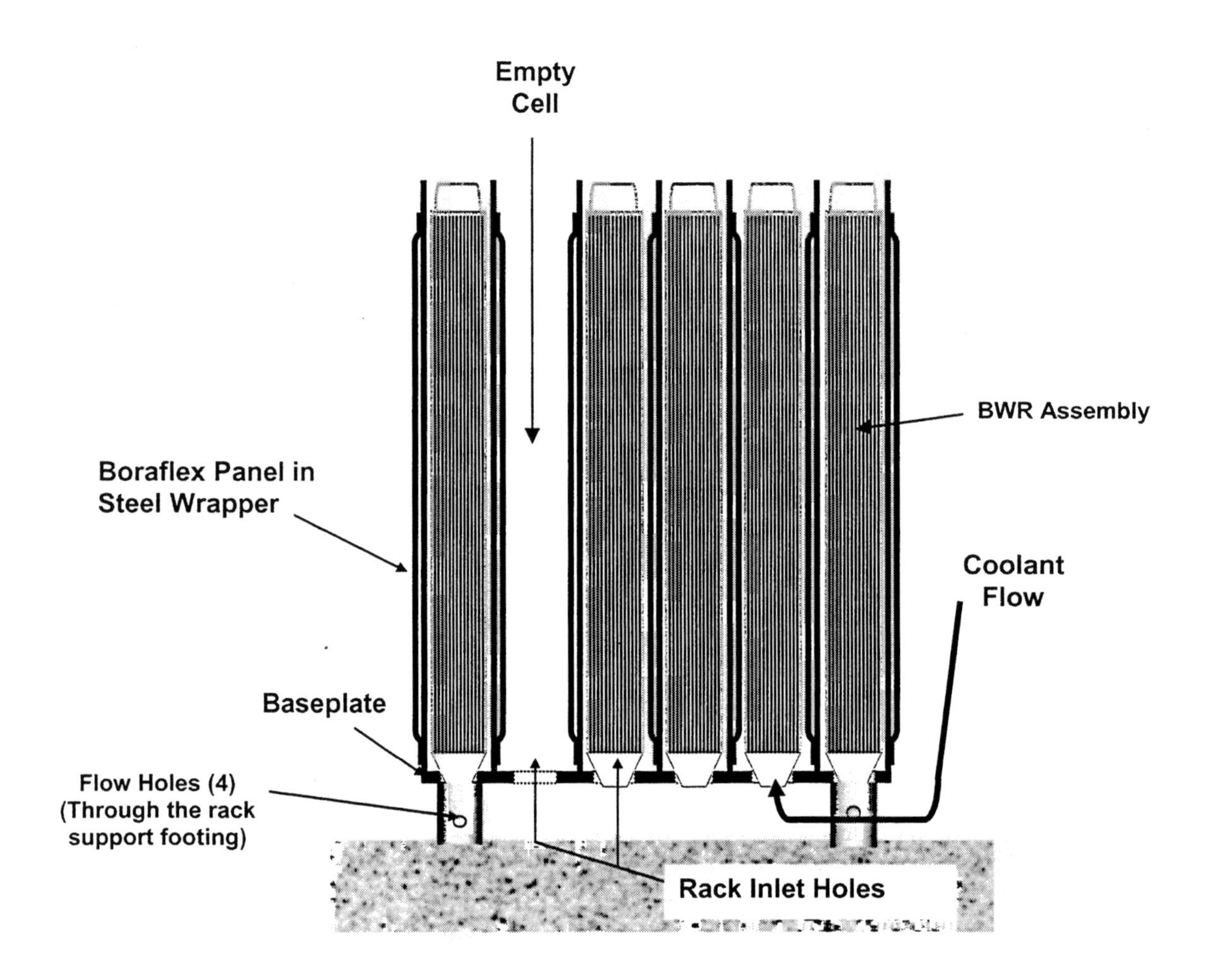

FIGURE 1.5 Dense spent fuel pool storage racks for BWR fuel. This cross-sectional illustration shows the principal elements of the spent fuel rack, which sits on the bottom of the pool. SOURCE: Nuclear Regulatory Commission briefing materials (2004).

2
TERRORIST ATTACKS ON SPENT FUEL STORAGE

This chapter addresses the final charge to the committee to "explicitly consider the risks of terrorist attacks on [spent fuel] and the risk these materials might be used to construct a radiological dispersal device." The concept of *risk* as applied to terrorist attacks underpins the entire statement of task for this study. Therefore, the committee addresses this final charge first to provide the basis for addressing the remainder of the task statement.

The chapter is organized into the following sections:

- Background on risk.
- Terrorist attack scenarios.
- Risks of terrorist attacks on spent fuel storage facilities.
- Findings and recommendations.

2.1 BACKGROUND ON RISK

"Risk" is a function of three factors (Kaplan and Garrick, 1981):

- The *scenario* describing the undesirable event.
- The *probability* that the scenario will occur.
- The *consequences* if the scenario should occur.

In the context of the present report, a *scenario* describes the modes and mechanisms of a possible terrorist attack against a spent fuel storage facility. For example, a scenario might involve a suicide attack with a hijacked civilian airliner. Another might involve a ground assault with a truck bomb. Several such scenarios are described later in this chapter and discussed in more detail in the committee's classified report.

Probability is a dimensionless quantity that expresses the likelihood that a given scenario will occur over a specified time period. If the occurrence of a scenario is judged to be impossible, it would have a probability of 0.0. On the other hand, if the scenario were judged to be certain, it has a probability of 1.0. A scenario that had a 50 percent chance of occurrence during the period contemplated would have a probability of 0.5.

Consequences describe the undesirable results if the scenario were to occur. For example, a terrorist attack on a spent fuel storage facility could release ionizing radiation to the environment.[1] The exposure of the public to this radiation could have both deterministic and stochastic effects. The former would occur from short-term exposures to very high doses of ionizing radiation, the latter to smaller doses that might have no immediate effects

[1] Terrorist scenarios and consequences are being described here for the sake of illustration. One should not conclude from this description that the committee believes that such consequences would necessarily occur as the result of a terrorist attack on a spent fuel storage facility.

but could result in cancer induction some years or decades later.[2] Consequences also could be described in terms of economic damage. These could arise, for example, from the loss of use of the facility and surrounding areas or costs to clean up those areas. There also could be severe psychological consequences that could drive changes in public acceptance of commercial nuclear energy.

The quantitative expression for the risk of a particular scenario, for example a suicide terrorist attack with a hijacked airliner, is

$$\text{Risk}_{\text{airliner attack}} = \text{Probability}_{\text{airliner attack}} \times \text{Consequences}_{\text{airliner attack}} \quad (1)$$

The total risk would be the sum of the risks for all possible independent attack scenarios. For example, if a spent fuel storage facility was determined to be vulnerable to attacks using airliners, truck bombs, and armed assaults, the total risk would be calculated as

$$\text{Risk}_{\text{total}} = \text{Risk}_{\text{airliner attack}} + \text{Risk}_{\text{truck bomb attack}} + \text{Risk}_{\text{armed assault attack}} \quad (2)$$

Such equations are routinely used to calculate the risks of various industrial accidents, including accidents at nuclear power plants, through a process known as *probabilistic risk assessment*. Each accident is assigned a numerical probability based on a careful analysis of the sequence of failures (e.g., human or mechanical failures) that could produce the accident. The consequences of such accidents are typically expressed in terms of injuries, deaths, or economic losses.

It is possible to estimate the risks of industrial accidents because there are sufficient experience and data to quantify the probabilities and consequences. This is not the case for terrorist attacks. To date, experts have not found a way to apply these quantitative risk equations to terrorist attacks because of two primary difficulties: The first is to develop a complete set of bounding scenarios for such attacks; the second is to estimate their probabilities. These depend on impossible-to-quantify factors such as terrorist motivations, expertise, and access to technical means.[3] They also depend on the effectiveness of measures that might prevent or mitigate such attacks.

In the absence of quantitative information on risks, one could attempt to make qualitative risk comparisons. Such comparisons could estimate, for example, the relative risks of attacks on spent fuel storage facilities versus attacks on commercial nuclear power reactors or other critical infrastructure such as chemical plants. Although a comparison of such risks is beyond the scope of this study, the committee recognizes that policy decisions about spent fuel storage may need to take into account such comparative risk issues,

---

[2] Such cancers would likely not be directly traceable to the radiation dose received from a terrorist attack and would likely be indistinguishable from the large population of cancers that result from other causes.

[3] Political scientists and counter-terror specialists have argued whether terrorists seek headlines, casualties, or both (e.g., Jenkins 1975, 1985). The September 11, 2001, attacks in the United States and the March 11, 2004, attacks in Spain demonstrate that some terrorists, particularly those of al-Qaida and its allies, intend to commit mass murder and/or mass economic disruption, both of which may have important political consequences. Further information about the motivation of terrorists is provided in NRC (2002).

especially for decisions regarding the expenditure of limited societal resources to address terrorist threats.

The 2002 National Research Council report *Making the Nation Safer: The Role of Science and Technology in Countering Terrorism* framed this issue as follows (NRC, 2002, p. 43):

> The potential vulnerabilities of NPPs [nuclear power plants] to terrorist attack seem to have captured the imagination of the public and the media, perhaps because of a perception that a successful attack could harm large populations and have severe economic and environmental consequences. There are, however, many other types of large industrial facilities that are potentially vulnerable to attack, for example, petroleum refineries, chemical plants, and oil and liquefied natural gas supertankers. These facilities do not have the robust construction and security features characteristic of NPPs, and many are located near highly populated urban areas.

Groups seeking to carry out high-impact terrorism will likely choose targets that have a high probability of being attacked successfully.[4] If success is measured by the number of people killed and injured or the permanent destruction of property, then spent fuel storage facilities may not make good terrorist targets owing to their relatively robust construction (see Chapters 1 and 3) and security. Industrialized societies like the United States provide terrorists a large number of "soft" (i.e., unprotected) targets that could be attacked more easily with greater effect than spent fuel storage facilities. These include chemical plants, refineries, transportation systems, and other facilities where large numbers of people gather (see NRC, 2002).

On the other hand, there are other success criteria that might influence a terrorist's decision to attack a "hard" (i.e., robust or well protected) target such as a commercial nuclear power plant and its spent fuel storage facilities. Such attacks could spread panic and shut down the power plant for an extended period of time even with no loss of life. Moreover, an attack that resulted in the release of radioactive material could threaten the viability of commercial nuclear power.

These considerations led the committee to conclude that it could not address its charge using quantitative and comparative risk assessments. The committee decided instead to examine a range of possible terrorist attack scenarios in terms of (1) their potential for damaging spent fuel pools and dry storage casks; and (2) their potential for radioactive material releases. This allowed the committee to make qualitative judgments about the vulnerability of spent fuel storage facilities to terrorist attacks and potential measures that could be taken to mitigate them.

---

[4] This point was made to the committee in a briefing by the Department of Homeland Security, where "success" means that the terrorist was able to achieve the goals of the attack, whatever they might be.

2.2 TERRORIST ATTACK SCENARIOS

It is possible to imagine a wide range of terrorist attacks against spent fuel storage facilities. Each would have a range of potential consequences depending on the characteristics of the attack and the facility being targeted as well as any post-attack mitigative actions to prevent or reduce the release of radioactive material. The committee focused its discussions about terrorist attacks around the concept of a *maximum credible scenario*—that is, an attack that is physically possible to carry out and that produces the most serious potential consequences within a given class of attack scenarios.

The following example illustrates the concept: One of the scenario classes considered by the committee in this chapter involves suicide attacks against spent fuel storage facilities with civilian passenger aircraft. The physics of such attacks are well understood: In general, heavier and higher-speed aircraft produce greater impact forces than lighter and slower aircraft, all else being equal. Consequently, the maximum credible scenario for suicide attacks involving civilian passenger aircraft would utilize the largest civilian passenger aircraft widely used in the United States flying at maximum cruising speed and hitting the facility at its most vulnerable point. Such an attack provides an upper bound to the damage that could be inflicted by this type of aircraft attack.

The maximum credible scenario is particularly useful for obtaining a general understanding of the damage that could be inflicted, but it would not necessarily apply to every spent fuel storage facility. To be judged a "credible" scenario, the terrorist must be able to successfully carry it out as designed—for example, to hit a spent fuel storage facility with the largest civilian aircraft at its most vulnerable point. This would rule out attacks that are physically impossible, such as flying a large civilian aircraft into a facility that is located below ground level or protected by surrounding hills or buildings. This also would rule out attacks involving weapons that are not available to terrorists (e.g., aircraft-launched weapons such as "bunker-buster" bombs or nuclear weapons).

This is not intended, however, to rule out attacks that are judged to have a low probability for success simply because terrorists might lack the skill and knowledge or luck to carry them out. In fact, if the consequences of such attacks were severe, policy makers might still decide that prudent mitigating actions should be taken regardless of their low probabilities of occurrence.[5] This might be especially true if quick, inexpensive fixes could be implemented. The main benefit of analyzing the maximum credible scenario is that it provides decision makers with a better characterization of the full range of potential consequences so that sound policy judgments can be made.

The analyses carried out for the Nuclear Regulatory Commission (described in the committee's classified report) do not consider maximum credible scenarios. Instead, the analyses employ *reference scenarios* that are based either on the characteristics of previous terrorist attacks or on qualitative judgments of the technical means and methods that might be employed in attacks against spent fuel storage facilities. Although such reference scenarios are useful for gaining insights on potential consequences of terrorist attacks, they

---

[5] The Department of Energy, for example, routinely examines the consequences of very low probability events involving nuclear weapons safety and security; see, for example, AL 56XB Development and Production Manual published by the U.S. Department of Energy, National Nuclear Security Administration. See *http://prp.lanl.gov/documents/d_p_manual.asp*.

are not necessarily bounding. This becomes important when the reference scenario attack results in damage to a facility that verges on failure.

The committee prefers a maximum credible scenario approach for one important reason: It believes that terrorists who choose to attack hardened facilities like spent fuel storage facilities would choose weapons capable of producing maximum destruction. **Of course, once the consequences of such attacks are known, an element of expert judgment is required to determine whether such attacks have a high likelihood of being carried out as designed. Such judgment is especially important when making policy decisions about actions to reduce the vulnerabilities of facilities to such attacks.**

The consequences of terrorist attacks can be described in terms of either *maximum credible releases* or *best-estimate releases*. The former describes the largest releases of radioactive material following an attack based on quantitative analytical models (e.g., the MELCOR computer code described in Chapter 3). The latter describes the median estimates from such models. In both cases, the estimates may not account for mitigative actions that could be taken after an attack to reduce or even eliminate releases. The Nuclear Regulatory Commission analyses reviewed by the committee in its classified report are best-estimate releases for various terrorist attack scenarios. The estimates in NUREG-1738 (USNRC, 2001a) and Alvarez et al. (2003a), on the other hand, describe maximum-credible to worst-case releases.[6]

The committee considered four classes of terrorist attack scenarios in this study:

- Air attacks using large civilian aircraft or smaller aircraft laden with explosives.
- Ground attacks by groups of well-armed and well-trained individuals.
- Attacks involving combined air and land assaults.
- Thefts of spent fuel for use by terrorists (including knowledgeable insiders) in radiological dispersal devices.

The committee devoted time at its meetings discussing these scenarios. It also received briefings on possible scenarios from Nuclear Regulatory Commission staff and suggestions for scenarios from the Department of Homeland Security (DHS), other experts, and the public. Some scenarios were dismissed by the committee as not credible. An example of such a scenario is an attack on a spent fuel storage facility with a nuclear weapon. Such weapons would be relatively difficult[7] for terrorists to build or steal. Even if such a weapon could be obtained, the committee can think of no reason that it would be used against a spent fuel storage facility rather than another target. There are easier ways to attack spent fuel storage facilities, as discussed in the classified report, and there are more attractive targets for nuclear weapons, for example, large population centers.

---

[6] Worst-case releases are based on the most unfavorable conditions that could occur in a given scenario, regardless of whether those conditions were physically realistic. For example, a worst-case estimate of the radionuclide releases from an attack on a spent fuel pool might assume that all of the volatile radionuclides contained in the spent fuel would be released, even if quantitative analytical models showed that such releases were very unlikely to occur.

[7] Difficult but certainly not impossible. See Chapter 2 in NRC (2002).

Given the experience of September 11, 2001, and the attacks that have occurred in other parts of the world, it is clear to the committee that the ability of the most capable terrorists to carry out attacks is limited only by their access to technical means. It is probably not limited by the ability of terrorist organizations to recruit or train attackers or bring them and any needed equipment into the United States—if indeed they are not already here. Moreover, the demonstrated willingness of terrorists to carry out suicide attacks greatly expands the scenarios that need to be considered when analyzing potential threats.

As is discussed in some detail in Chapters 3 and 4, the facilities used to store spent fuel at nuclear power plants are very robust. Thus, only attacks that involve the application of large energy impulses or that allow terrorists to gain interior access have any chance of releasing substantial quantities of radioactive material. This further restricts the scenarios that need to be considered. For example, attacks using rocket-propelled grenades (RPGs) of the type that have been carried out in Iraq against U.S. and coalition forces would not likely be successful if the intent of the attack is to cause substantial damage to the facility. Of course, such an attack would get the public's attention and might even have economic consequences for the attacked plant and possibly the entire commercial nuclear power industry.

The threat scenarios summarized in this chapter are based on documents provided to the committee, briefings received at committee meetings, and the committee's own expert judgment.[8] Further overview and information on nuclear and radiological threats in general can be found in the NRC (2002) report and references therein.

2.2.1 Air Attacks

The September 11, 2001, attacks[9] demonstrated that terrorists are capable of successfully attacking fixed infrastructure with large civilian jetliners. The security of civilian passenger airliners has been improved since these attacks were carried out, and the vulnerability of civilian passenger aircraft to highjacking has been reduced. Nevertheless, the committee judges, based on the evidence made available to it during this study, that attacks with civilian aircraft remain a credible threat. Such aircraft are used routinely in freight and charter services, and large numbers of such aircraft enter the United States from other countries each day. Improvements to ground security or cargo inspection would likely not eliminate the threat posed by an air crew willing to stage a suicide attack with a chartered air freighter.

Although the September 11, 2001, attacks utilized Boeing 757 and 767 airliners, larger aircraft (Boeing 747, 777; Airbus 340) are in routine use around the world, and an even larger aircraft (Airbus 380) is entering production. Assaults by such large aircraft could impart enormous energy impulses to spent fuel storage facilities. Additionally, attacks with

---

[8] The committee found limited information in the open literature on various scenarios for terrorist attacks on nuclear plants and their spent fuel storage facilities.

[9] The al-Qaida terrorist organization hijacked and crashed two Boeing 767 airliners into Towers 1 and 2 of the World Trade Center building in New York and a Boeing 757 airliner into the Pentagon building in Arlington, Virginia. A second Boeing 757, which was believed to be targeted either on the White House or the U.S. Capitol (see National Commission on Terrorist Attacks Upon the United States, Staff Statement No. 16 [Outline of the 9/11 Plot], pages 18-19) crashed in an open field near Jennerstown, Pennsylvania.

aircraft carrying large fuel loads could produce fires that would greatly complicate rescue and recovery efforts.

Previous studies on aircraft crash impacts (Droste et al., 2002; Lange et al., 2002; HSK, 2003; RBR Consultants, 2003; Thomauske, 2003) suggest that the consequences of a heavy aircraft crash on a nuclear installation depend on factors such as the following:

- Type and design of the aircraft.
- Speed of the aircraft.
- Fuel loading of the aircraft and total weight at impact.
- Angle-of-attack and point-of-impact on the facility.
- Construction of the facility.
- Location of the target with respect to ground level (i.e., below or above grade).[10]
- The presence of surrounding buildings and other obstacles (e.g., hills, transmission lines) that might block certain potential flight paths into the facility.

In other words, the consequences of such attacks are scenario- and plant-design specific. It is not possible to make any general statements about spent fuel storage facility vulnerabilities to air attacks that would apply to all U.S. commercial nuclear power plants.

U.S. commercial nuclear power plants are not required by the Nuclear Regulatory Commission to defend against air attacks. The Commission believes that it is the responsibility of the U.S. government to implement security measures to prevent such attacks. The commercial nuclear industry shares this view. The Nuclear Regulatory Commission staff informed the committee that the Commission has directed power plant operators to take steps to reduce the likelihood of serious consequences should such attacks occur. The staff also informed the committee that the Commission may issue additional directives once the vulnerability analyses it is sponsoring at Sandia National Laboratories are completed. These analyses are described in the committee's classified report (see also Chapters 3 and 4 in this report).

2.2.2 Ground Attacks

Ground attacks on a nuclear facility could take three forms: (1) a direct assault on the facility by armed groups, (2) a stand-off attack using appropriate weapons, or (3) an assault having both air and ground components. The direct assault would likely be carried out by a group of well-armed and trained attackers, perhaps working with the assistance of an insider. The objective of such an attack would likely be to gain entry to protected and vital areas of the plant (FIGURE 2.1) to carry out radiological sabotage. The attackers would need to have knowledge of the design, location, and operation of the spent fuel facility to carry out such an attack successfully.

Commercial nuclear power plants are required by the Nuclear Regulatory Commission to maintain a professional guard force at each plant to defend against a Commission-developed design basis threat (DBT), which includes a ground assault. The protective force is a critical part of a nuclear power plant's security system for deterring,

---

[10] All current dry cask storage facilities in the United States are constructed at ground level, whereas spent fuel pools can be located above or below grade, depending on plant design (see Chapter 3).

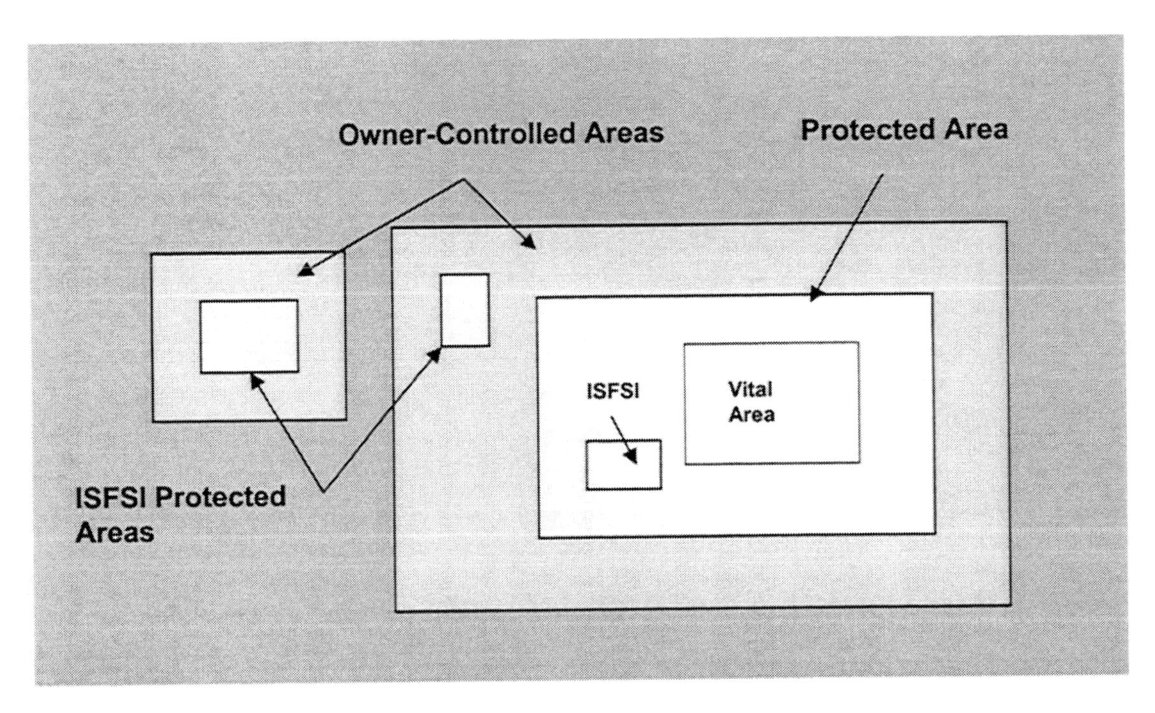

FIGURE 2.1 Commercial nuclear power plant sites are demarcated as shown for security purposes. The part of the power plant site over which the plant operator exercises control is referred to as the *owner-controlled area*. This usually corresponds to the boundary of the site. Located within this area are one or more *protected areas* to which access is restricted using guards, fences, and other barriers. Dry cask storage facilities, formally referred to as Independent Spent Fuel Storage Installations (ISFSIs), are located within these areas. The *vital area* of the plant contains the reactor core, support buildings, and the spent fuel pool. It is the most carefully controlled and guarded part of the plant site. SOURCE: Modified from Nuclear Regulatory Commission briefing materials (2004).

detecting, thwarting, or impeding attacks. The Commission staff declined to provide a formal briefing to the committee on the DBT for radiological sabotage, asserting that the committee did not have a need to know this information. Nevertheless, the committee was able to discern the details of the DBT from a series of presentations made by Nuclear Regulatory Commission staff. Commission staff also provided a fact check of this information as the classified report was being finalized.

Power plant operators are required to demonstrate to the Commission's satisfaction that there is "high assurance" that their guard forces can thwart the Commission-defined DBT assault. This guard force also must be able to provide deterrence against a beyond-DBT attack depending on the adversarial force. Reinforcing forces would be provided by local and state law enforcement as well as federal forces. The Commission staff also informed the committee that since the September 11, 2001, attacks, the Commission has been working with DHS to improve coordination procedures with federal, state, and local agencies to improve their response capabilities in the event of an attack. DHS also is making grants to local law enforcement agencies around power plant sites to raise their capabilities to respond to requests for assistance.

Since the September 11, 2001, attacks, the Nuclear Regulatory Commission has issued directives to power plant operators to enhance protection against vehicle bombs. The Commission also has issued directives to power plant operators to enhance protection against insider threats.

The committee does not have enough information to judge whether the measures at power plants are in fact sufficient to defend against either a DBT or a beyond-DBT attack on spent fuel storage. The Nuclear Regulatory Commission declined to provide detailed briefings to the committee on surveillance, security procedures, and security training at commercial nuclear power plants. Consequently, the committee was unable to evaluate their effectiveness. A recent General Accounting Office report (GAO, 2003) was critical of some of these procedures, but the committee has no basis for judging whether these criticisms were justified. Nevertheless, the committee judges that surveillance and security procedures at commercial nuclear power plants are just as important as physical barriers in preventing successful terrorist attacks and mitigating their consequences.

2.2.3 Attacks Having Both Air and Ground Components

Hybrid attacks that combine aspects of both air and ground attacks also could be mounted by terrorists. These could deliver attacking forces directly to a spent fuel storage facility, bypassing the security perimeters and security personnel deployed to protect against a ground attack. The committee considered various scenarios for such attacks. The committee judges that some scenarios are feasible. Details are provided in the classified report.

2.2.4 Terrorist Theft of Spent Fuel for Use in a Radiological Dispersal Device (RDD)

An RDD, or so-called dirty bomb, is a device that disperses radioactive material using chemical explosives or other means (NRC, 2002). RDDs do not involve fission-induced explosions of the kind associated with nuclear weapons. While RDD attacks can be carried out with any source of radioactivity, this discussion is confined to scenarios that involve the theft of spent fuel for such use.[11] A crude RDD device could be fabricated simply by loading stolen spent fuel onto a truck carrying high explosives. The truck could be driven to another location and detonated. The dispersal of radioactivity from such an attack would be unlikely to cause many immediate deaths, but there could be fatalities from the chemical explosion as well as considerable cleanup costs and adverse psychological effects.

It would be difficult for terrorists to steal a large quantity of spent fuel (e.g., a single spent fuel assembly) for use in an RDD for three reasons. First, spent fuel is highly radioactive and therefore requires heavy shielding to handle. Second, the use of heavy equipment would be required to remove spent fuel assemblies from a pool or dry cask. Third, controls are in place at plants to deter and detect such thefts. Additional details on these controls are provided in the classified report.

Theft and removal of an assembly or individual fuel rods during an assault on the plant might be easier, because the guard force would likely be preoccupied defending the plant. However, the amount of material that could be removed would be small, and getting it

---

[11] An attack on a spent fuel facility that resulted in the direct release of radioactivity would be an act of radiological sabotage of the kind considered previously in this chapter.

out of the plant would be time consuming and obvious to the plant defenders and other responding forces.

There are broken fuel rods and other debris, mostly from older assemblies, in storage at many plants. These materials are typically stored along the sides of the spent fuel pools and could be more easily removed from the plant than an entire assembly. Pieces of fuel rods also are sometimes intentionally removed from assemblies for offsite laboratory analysis. Some plants have misplaced fuel rod pieces.[12] A knowledgeable insider might be able to retrieve some of this material from the pool, but getting it out of the plant under normal operating conditions would be difficult.

Even the successful theft of a part of a spent fuel rod would provide a terrorist with only a relatively small amount of radioactive material. Superior materials could be obtained from other facilities. This material also can be purchased (Zimmerman and Loeb, 2004).

Moreover, even with explosive dissemination, it is unlikely that much of the spent fuel will be aerosolized unless it is incorporated into a well-designed RDD. More likely, such an event would break up and scatter the fuel pellets in relatively large chunks, which would not pose an overwhelming cleanup challenge.

Even though the likelihood of spent fuel theft appears to be small, it is nevertheless important that the protection of these materials be maintained and improved as vulnerabilities are identified.

2.3 RISKS OF TERRORIST ATTACKS ON SPENT FUEL STORAGE FACILITIES

Nuclear Regulatory Commission staff told the committee that it believes that the consequences of a terrorist attack on a spent fuel pool would likely unfold slowly enough that there would be time to take mitigative actions to prevent a large release of radioactivity. They also pointed out that since the September 11, 2001, attacks, the Nuclear Regulatory Commission has issued several orders that contain Interim Compensatory Measures that require power plant operators to consider potential mitigative actions in the event of such an attack. The committee received a briefing on some of these measures at one of its meetings. According to Commission staff, such measures provide an additional margin of safety.

The nuclear industry and the Nuclear Regulatory Commission have also asserted that the robust construction and stringent security requirements at nuclear power plants[13] make them less vulnerable to terrorist attack than softer targets such as chemical plants and refineries (e.g., Chapin et al., 2002). They argue that scarce resources should be devoted to

---

[12] For example, at the Millstone and Vermont Yankee plants in 2000 and 2003, respectively. In the case of Millstone, the Nuclear Regulatory Commission determined on the basis of extensive analysis that these rods were likely disposed of as low-level waste. After the committee's classified report was published, Commission staff informed the committee that Vermont Yankee had accounted for the missing rod segments and that Humbolt Bay had uncovered and is investigating an inventory discrepancy involving spent fuel rod segments.

[13] These arguments tend to be generic in nature and do not differentiate spent fuel pools from the rest of the power plant.

upgrading security at these other critical facilities rather than at already well-protected nuclear plants.

There are two unstated propositions in the argument that nuclear plants are less vulnerable than other facilities. The first speaks to the probability of terrorist attacks on such facilities; the second speaks to the consequences:

- *Proposition 1:* Nuclear power plants (and their spent fuel facilities) are less desirable as terrorist targets because they are robust and well protected.
- *Proposition 2:* If attacked, nuclear plants (and their spent fuel storage facilities) are likely to sustain little or no damage because they are robust and well protected.

The committee obtained a briefing from the Department of Homeland Security to address the first proposition. Details are provided in the classified report.

While the committee's classified report was in review, the National Commission on Terrorist Attacks Upon the United States issued a staff paper (Staff Statement No. 16, Outline of the 9/11 Plot, pages 12-13) suggesting that al-Qaida initially included unidentified nuclear plants among an expanded list of targets for the September 11, 2001, attacks. According to that report, these plants were eliminated from the target list along with several other facilities when the terrorist organization scaled back the number of planned attacks. Nevertheless, if this information is correct, it provides further indications that commercial nuclear power plants are of interest to terrorist groups,[14] even though softer targets may have a higher priority with many terrorists.

With respect to the first proposition, the committee judges that it is not prudent to dismiss nuclear plants, including their spent fuel storage facilities, as undesirable targets for attacks by terrorists.

As to the second proposition that terrorist attacks are likely to cause little or no damage, a poorly designed attack or an attack by unsophisticated terrorists might produce little physical damage to the plant. There could, however, be severe adverse psychological effects from such an attack that could have considerable economic consequences. On the other hand, attacks by knowledgeable terrorists with access to advanced weapons might cause considerable physical damage to a spent fuel storage facility, especially in a suicide attack.

It is important to recognize that an attack that damages a power plant or its spent fuel facilities would not necessarily result in the release of *any* radioactivity to the environment. While it may not be possible to deter such an attack, there are many potential mitigation steps that can be taken to lower its potential consequences should an attack occur. These are discussed in some detail in the committee's classified report (see also Chapters 3 and 4 in this report).

---

[14] In another example of concern, police in Toronto, Canada, detained 19 men in August 2003 based on suspicious activities that included surveillance and flying lessons that would take them over a nuclear power plant (Ferguson et al., 2004).

In summary, the committee judges that the plausibility of an attack on a spent fuel storage facility, coupled with the public fear associated with radioactivity, indicates that the possibility of attacks cannot be dismissed.

2.4 FINDINGS AND RECOMMENDATIONS

With respect to the committee's task to "explicitly consider the risks of terrorist attacks on [spent fuel] and the risk these materials might be used to construct a radiological dispersal device," the committee offers the following findings and recommendations:

FINDING 2A: The probability of terrorist attacks on spent fuel storage cannot be assessed quantitatively or comparatively. Spent fuel storage facilities cannot be dismissed as targets for such attacks because it is not possible to predict the behavior and motivations of terrorists, and because of the attractiveness of spent fuel as a terrorist target given the well-known public dread of radiation.

Terrorists view nuclear power plant facilities as desirable targets because of the large inventories of radionuclides they contain. The committee believes that knowledgeable terrorists might choose to attack spent fuel pools because (1) at U.S. commercial power plants, these pools are less well protected structurally than reactor cores; and (2) they typically contain inventories of medium- and long-lived radionuclides that are several times greater than those contained in individual reactor cores.

FINDING 2B: The committee judges that the likelihood terrorists could steal enough spent fuel for use in a significant radiological dispersal device is small.

Spent fuel assemblies in pools or dry casks are large, heavy, and highly radioactive. They are too large and radioactive to be handled by a single individual. Removal of an assembly from the pool or dry cask would prove extremely difficult under almost any terrorist attack scenario. Attempts by a knowledgeable insider(s) to remove single rods and related debris from the pool might prove easier, but it would likely be very difficult to get it out of the plant under normal operating conditions. Theft and removal during an assault on the plant might be easier because the guard force would likely be occupied defending the plant. However, the amount of material that could be removed would be small. Moreover, there are other facilities from which highly radioactive material could be more easily stolen, and this material also can be purchased. Even though the likelihood of spent fuel theft appears to be small, it is nevertheless important that the protection of these materials be maintained and improved as vulnerabilities are identified.

> **RECOMMENDATION: The Nuclear Regulatory Commission should review and upgrade, where necessary, its security requirements for protecting spent fuel rods not contained in fuel assemblies from theft by knowledgeable insiders, especially in facilities where individual fuel rods or portions of rods are being stored in pools.**

FINDING 2C: A number of security improvements at nuclear power plants have been instituted since the events of September 11, 2001. The Nuclear Regulatory Commission did not provide the committee with enough information to evaluate the effectiveness of these procedures for protecting stored spent fuel.

Surveillance and security procedures are just as important as physical barriers in preventing and mitigating terrorist attacks. The Nuclear Regulatory Commission declined to provide the committee with detailed briefings on the surveillance and security procedures that are now in place to protect spent fuel facilities at commercial nuclear power plants against terrorist attacks. Although the committee did learn about some of the changes that have been instituted since the September 11, 2001, attacks, it was not provided with enough information to evaluate the effectiveness of procedures now in place.

RECOMMENDATION: Although the committee did not specifically investigate the effectiveness and adequacy of improved surveillance and security measures for protecting stored spent fuel, an assessment of current measures should be performed by an independent[15] organization.

[15] That is, independent of the Nuclear Regulatory Commission and the nuclear industry.

3

SPENT FUEL POOL STORAGE

This chapter addresses the first charge of the committee's statement of task to assess "potential safety and security risks of spent nuclear fuel presently stored in cooling pools at commercial reactor sites."[1] As noted in Chapter 1, storage of spent fuel in pools at commercial reactor sites has three primary objectives:

- Cool the fuel to prevent heat-up to high temperatures from radioactive decay.
- Shield workers and the public from the radiation emitted by radioactive decay in the spent fuel and provide a barrier for any releases of radioactivity.
- Prevent criticality accidents.

The first two of these objectives could be compromised by a terrorist attack that partially or completely drains the spent fuel pool.[2] The committee will refer to such scenarios as "loss-of-pool-coolant" events. Such events could have several deleterious consequences: Most immediately, ionizing radiation levels in the spent fuel building rise as the water level in the pool falls. Once the water level drops to within a few feet (a meter or so) of the tops of the fuel racks, elevated radiation fields could prevent direct access to the immediate areas around the lip of the spent fuel pool building by workers. This might hamper but would not necessarily prevent the application of mitigative measures, such as deployment of fire hoses to replenish the water in the pool.

The ability to remove decay heat from the spent fuel also would be reduced as the water level drops, especially when it drops below the tops of the fuel assemblies. This would cause temperatures in the fuel assemblies to rise, accelerating the oxidation of the zirconium alloy (zircaloy) cladding that encases the uranium oxide pellets. This oxidation reaction can occur in the presence of both air and steam and is strongly exothermic—that is, the reaction releases large quantities of heat, which can further raise cladding temperatures. The steam reaction also generates large quantities of hydrogen:

Reaction in air:	$Zr + O_2 \rightarrow ZrO_2$	heat released = 1.2×10^7 joules/kilogram
Reaction in steam:	$Zr + 2H_2O \rightarrow ZrO_2 + 2H_2$	heat released = 5.8×10^6 joules/kilogram

[1] A basic description of pool storage can be found in Chapter 1 and historical background can be found in Appendix D. Section 3.1 provides additional technical details about pool storage.
[2] The committee could probably design configurations in which fuel might be deformed or relocated to enable its re-criticality, but the committee judges such an event to be unlikely. Also, the committee notes that while re-criticality would certainly be an undesirable outcome, criticality accidents have happened several times at locations around the world and have not been catastrophic offsite. An accompanying breach of the fuel cladding would still be the chief concern.

These oxidation reactions can become locally self-sustaining (i.e., autocatalytic[3]) at high temperatures (i.e., about a factor of 10 higher than the boiling point of water) if a supply of oxygen and/or steam is available to sustain the reactions. (These reactions will not occur when the spent fuel is under water because heat removal prevents such high temperatures from being reached).The result could be a runaway oxidation reaction—referred to in this report as a *zirconium cladding fire*—that proceeds as a burn front (e.g., as seen in a forest fire or a fireworks sparkler) along the axis of the fuel rod toward the source of oxidant (i.e., air or steam). The heat released from such fires can be even greater than the decay heat produced in newly discharged spent fuel.

As fuel rod temperatures increase, the gas pressure inside the fuel rod increases and eventually can cause the cladding to balloon out and rupture. At higher temperatures (around 1800°C [approximately 3300°F]), zirconium cladding reacts with the uranium oxide fuel to form a complex molten phase containing zirconium-uranium oxide. Beginning with the cladding rupture, these events would result in the release of radioactive fission gases and some of the fuel's radioactive material in the form of aerosols into the building that houses the spent fuel pool and possibly into the environment. If the heat from one burning assembly is not dissipated, the fire could spread to other spent fuel assemblies in the pool, producing a propagating zirconium cladding fire.

The high-temperature reaction of zirconium and steam has been described quantitatively since at least the early 1960s (e.g., Baker and Just, 1962). The accident at the Three Mile Island Unit 2 reactor and a set of experiments (e.g., CORA, FPT 1-6, CODEX, ORNL-VI, VERCORS) have provided a basis for understanding the phenomena of zirconium cladding fires and fission-product releases from irradiated fuel in a reactor core accident. This understanding and data from the experiments form the foundation for computer simulations of severe accidents involving nuclear fuel. These experiments and computer simulations are for inside-reactor vessel events rather than events in an open-air spent fuel pool array.

This chapter examines possible initiating factors for such loss-of-pool-coolant events and the potential consequences of such events. It is organized into the following four main sections:

- Background on spent fuel pool storage.
- Previous studies on safety and security of pool storage.
- Evaluation of the potential risks of pool storage.
- Findings and recommendations.

---

[3] That is, the reaction heat will increase temperatures in adjacent areas of the fuel rod, which in turn will accelerate oxidation and release even more heat. Autocatalytic oxidation leading to a "runaway" reaction requires a complex balance of heat and mass transfer, so assigning a specific ignition temperature is not possible. Empirical equations have been developed to predict the reaction rate as a function of temperature when steam and oxygen supply are not limited (see, e.g., Tong and Weisman, 1996, p. 223). Numerous scaled experiments have found that the oxidation reaction proceeds very slowly below approximately 900°C (1700°F).

3.1 BACKGROUND ON SPENT FUEL POOL STORAGE

After a power reactor is shut down, its nuclear fuel continues to produce heat from radioactive decay (see FIGURE 1.2). Although only one-third of the fuel in the reactor core is replaced during each refueling cycle, operators commonly offload the entire core (especially at pressurized water reactors [PWRs]) into the pool during refueling[4] to facilitate loading of fresh fuel or for inspection or repair of the reactor vessel and internals. Heat generation in the pool is at its highest point just after the full core has been offloaded.

Pool heat loads can be quite high, as exemplified by a "typical" boiling water reactor (BWR) which was used in some of the analyses discussed elsewhere in this chapter (this BWR is hereafter referred to as the "reference BWR"). This pool has approximately 3800 locations for storage of spent fuel assemblies, about 3000 of which are occupied by four-and-one-third reactor cores (13 one-third-core offloads) in a pool approximately 35 feet wide, 40 feet long, and 39 feet deep (10.7 meters wide, 12.2 meters long, and 11.9 meters deep) with a water capacity of almost 400,000 gallons (1.51 million liters). According to Nuclear Regulatory Commission staff, the total decay heat in the spent fuel pool is 3.9 megawatts (MW) ten days after a one-third-core offload. The vast majority of this heat is from decay in the newly discharged spent fuel. Heat loads would be substantially higher in spent fuel pools that contained a full-core offload.

Although spent fuel pools have a variety of designs, they share one common characteristic: Almost all spent fuel pools are located outside of the containment structure that holds the reactor pressure vessel.[5] In some reactor designs, the spent fuel pools are contained within the reactor building,[6] which is typically constructed of about 2 feet of reinforced concrete (see FIGURE 3.1). In other designs, however, one or more walls of the spent fuel pool may be located on the exterior wall of an auxiliary building that is located adjacent to the containment building (see FIGURE 3.2). As described in more detail below, some pools are built at or below grade, whereas others are located at the top of the reactor building.

The enclosing superstructures above the pool are typically steel, industrial-type buildings designed to house cranes that are used to move reactor components, spent fuel, and spent fuel casks. These superstructures above the pool are designed to resist damage from seismic loads but not from large tornado-borne missiles (e.g., cars and telephone poles), which would usually impact the superstructures at low angles (i.e., moving horizontally). In contrast, the typical spent fuel pool is robust. The pool walls and the external walls of the building housing the pool (these external walls may incorporate one or more pool walls in some plants) are designed for seismic stability and to resist horizontal

---

[4] A 1996 survey by the Nuclear Regulatory Commission (USNRC, 1996) found that the majority of commercial power reactors routinely offload their entire core to the spent fuel pool during refueling outages. The practice is more common among PWRs than BWRs, which tend to offload only that fuel that is to be replaced, but some BWRs do offload the full core. In response to a committee inquiry, an Energy Resources International staff member confirmed that this is still the case today.
[5] The exceptions in the United States are the Mark III BWRs, which have two pools, one of which is inside the containment. As discussed in Appendix C, spent fuel pools at German commercial nuclear power plants also are located inside reactor containment structures.
[6] A PWR containment structure is a large, domed building that houses the reactor pressure vessel, the steam generators, and other equipment. In a BWR, the containment structure houses less equipment, is located closer in to the pressure vessel, and sits inside a building called the reactor building, which also houses the spent fuel pool and safety-related equipment to support the reactor.

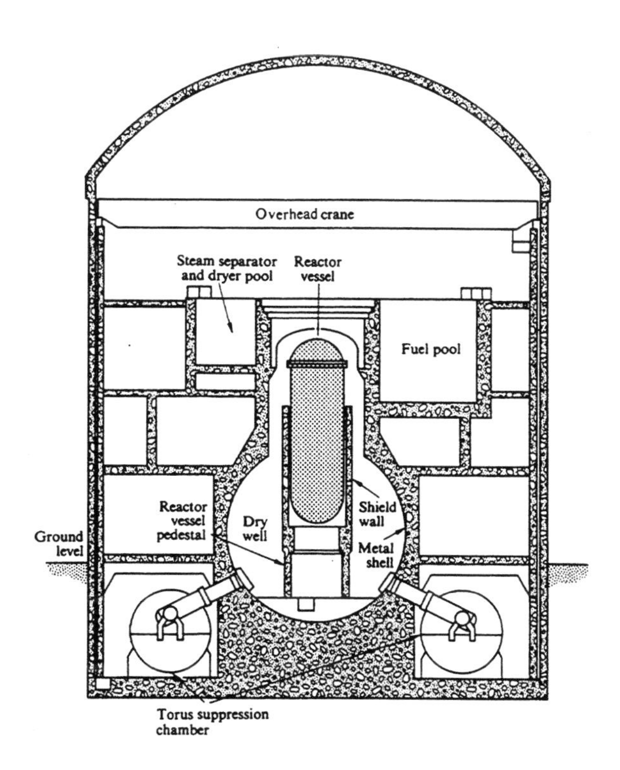

FIGURE 3.1 Schematic section through a G.E. Mark I BWR reactor plant. The spent fuel pool is located in the reactor building well above ground level. This diagram is for a BWR with a reinforced concrete superstructure (roof). Most designs have thin steel superstructures. SOURCE: Lamarsh (1975, Figure 11.3).

strikes of tornado missiles. The superstructures and pools were not, however, specifically designed to resist terrorist attacks.

The typical spent fuel pool is about 40 feet (12 meters) deep and can be 40 or more feet (12 meters) in each horizontal dimension. The pool walls are constructed of reinforced concrete typically having a thickness between 4 and 8 feet (1.2 to 2.4 meters). The pools contain a ¼- to ½-inch-thick (6 to 13 mm) stainless steel liner, which is attached to the walls with studs embedded in the concrete. The pools also contain vertical storage racks for holding spent and fresh fuel assemblies, and some pools have a gated compartment to hold a spent fuel storage cask while it is being loaded and sealed (see Chapter 4).

The storage racks are about 13 feet (4 meters) in height and are installed near the bottom of the spent fuel pool. The racks have feet to provide space between their bottoms and the pool floor. There is also space between the sides of the rack and the steel pool liners for circulation of water (FIGURE 3.3). There are about 26 feet (8 meters) of water above the top of the spent fuel racks. This provides substantial radiation shielding even when an assembly is being moved above the rack. Transfers of spent fuel from the reactor core to the spent fuel pool or from the pool to storage casks are carried out underwater to provide shielding and cooling.

The general elevation of the spent fuel pool matches that of the vessel containing the reactor core. Pressurized water reactor designs use comparatively shorter reactor

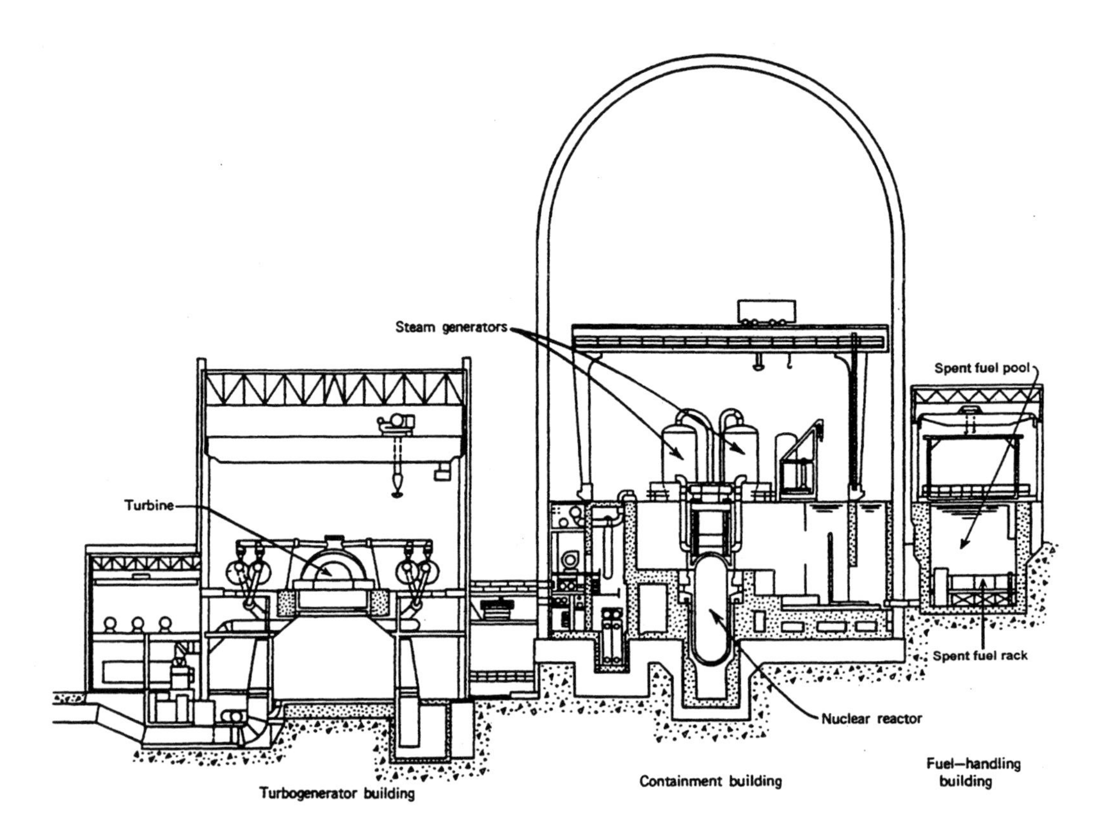

FIGURE 3.2 Schematic section through a PWR reactor plant. The spent fuel pool is located in the fuel-handling building next to the domed reactor containment building at or slightly below ground level. SOURCE: Modified from Duderstadt and Hamilton (1976, Figure 3-4).

vessels closer to ground level (grade) and also have spent fuel pools that are close to grade (FIGURE 3.2). The design shown in this figure is typical of the fuel pool arrangement for PWRs. Nuclear power plant sites that contain two reactors are usually arranged in a mirror-image fashion, with the two spent fuel pools (or a shared pool) located in a common area adjoining both reactor buildings. For single-plant or two-plant arrangements, the building covering the spent fuel pool and crane structures is typically an ordinary steel industrial building. There are 69 PWRs currently in operation in the United States; 6 PWRs have been decommissioned but continue to have active spent fuel pool storage.

In contrast, in boiling water reactor designs, the reactor vessel is at a higher elevation, and the BWR vessels are somewhat taller than PWR vessels.[7] Consequently, BWRs have more elevated spent fuel pools, generally well above grade. FIGURE 3.1 shows the general design for the 22 BWR Mark I plants operating in the United States.

Nuclear Regulatory Commission staff is conducting a survey of the plants to obtain a better understanding of the variations in design of spent fuel pools across the nation. The following information was provided to the committee from that survey:

---

[7] The higher elevation accommodates control mechanisms that sit under the reactor, and the extra height accommodates steam separation and drying equipment at the top of the vessel. The fuel is about the same length as PWR fuel.

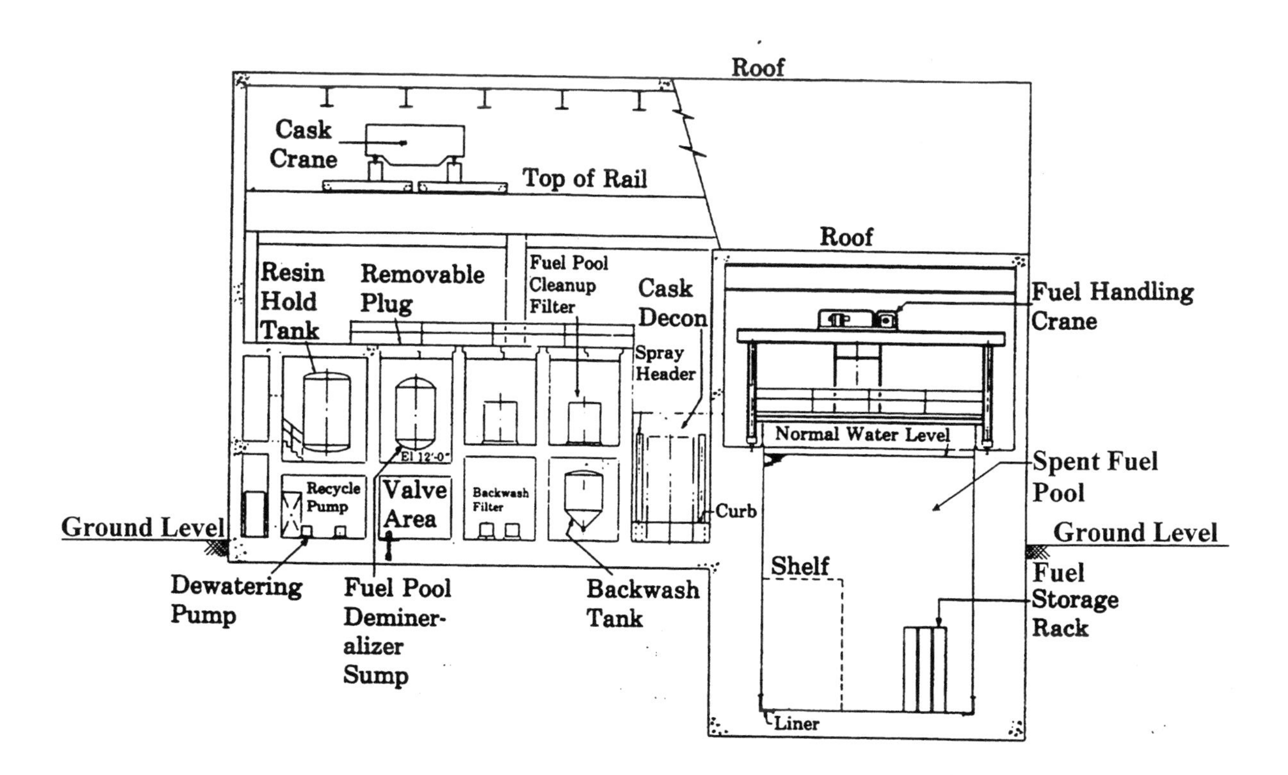

FIGURE 3.3 Example of a section of a PWR spent fuel pool and support facilities. The pool is located to the right in the figure; the support equipment to the left. SOURCE: American Nuclear Society (1988).

- PWR spent fuel pools: Spent fuel pools are located in buildings adjoining the reactor containment buildings at PWR plants (see FIGURE 3.2). Some pools are positioned such that their spent fuel is below grade. As shown in Figure 3.2, some pool walls also serve as the external walls of the spent fuel pool buildings. Some plants have structures surrounding the spent fuel pool building that would provide some shielding of the pools from low-angle line-of-sight attacks. A more complete plant survey would be needed to establish the extent of pool exposure to such attacks.
- BWR spent fuel pools: MARK I and II BWR plants are located above grade and are shielded by at least one exterior building wall. Some pools are also shielded by the reactor buildings. Some pools are also shielded by "significant" surrounding structures, and some have supplemental floor and column supports.

The vulnerability of a spent fuel pool to terrorist attack depends in part on its location with respect to ground level as well as its construction. Pools are potentially susceptible to attacks from above or from the sides depending on their elevation with respect to grade and the presence of surrounding shielding structures.

As noted in Chapter 1, nearly all pools contain high-density spent fuel racks. These racks allow approximately five times as many assemblies to be stored in the pool as would have been possible with the original racks, which had open lateral channels between the fuel assemblies to enhance water circulation.

3.2 PREVIOUS STUDIES ON SAFETY AND SECURITY OF POOL STORAGE

Several reports have been published on the safety of spent fuel pool storage. One of the earliest analyses was contained in the *Reactor Safety Study* (U.S. Atomic Energy Commission, 1975), which concluded that spent fuel pool safety risks were very much smaller than those involving the cores of nuclear reactors. This conclusion is not surprising: The cooling system in a spent fuel pool is simple. The coolant is at atmospheric pressure; the spent fuel is in a subcritical configuration and generates little heat relative to that generated in an operating reactor; and the design and location of piping in the pool make a severe loss-of-pool-coolant event unlikely during normal operating conditions. Despite changes in reactor and fuel storage operations, such as longer fuel residence times in the core and higher-density pool storage, the conclusions of that study are still broadly applicable today. It is important to recognize, however, that the *Reactor Safety Study* did not address the consequences of terrorist attacks.

The Nuclear Regulatory Commission and its contractors have periodically re-analyzed the safety of spent nuclear fuel storage (see Benjamin et al., 1979; BNL, 1987, 1997; USNRC, 1983, 2001a, 2003b). All of these studies suggest that a loss-of-pool-coolant event could trigger a zirconium cladding fire in the exposed spent fuel. The Nuclear Regulatory Commission considered such an accident to be so unlikely that no specific action was warranted, despite changes in reactor operations that have resulted in increased fuel burn-ups and fuel storage operations that have resulted in more densely packed spent fuel pools.

In 2001, the Nuclear Regulatory Commission published NUREG-1738, *Technical Study of Spent Fuel Pool Accident Risk at Decommissioning Nuclear Power Plants,* to provide a technical basis for rulemaking for power plant decommissioning (USNRC, 2001a). A draft of the study was issued for public comments, including comments by the Advisory Committee on Reactor Safeguards and a quality review of the methods, assumptions, and models used in the analysis was carried out by the Idaho National Engineering and Environmental Laboratory.

The study provided a probabilistic risk assessment that identified severe accident scenarios and estimated their consequences. The analysis determined, for a given set of fuel characteristics, how much time would be required to boil off enough water to allow the fuel rods to reach temperatures sufficient to initiate a zirconium cladding fire.

The analysis suggested that large earthquakes and drops of fuel casks from an overhead crane during transfer operations were the two event initiators that could lead to a loss-of-pool-coolant accident. For cases where active cooling (but not the coolant) has been lost, the thermal-hydraulic analyses suggested that operators would have about 100 hours (more than four days) to act before the fuel was uncovered sufficiently through boiling of cooling water in the pool to allow the fuel rods to ignite. This time was characterized as an "underestimate" given the simplifications assumed for the loss-of-pool-coolant scenario.

The overall conclusion of the study was that the risk of a spent fuel pool accident leading to a zirconium cladding fire was low despite the large consequences because the predicted frequency of such accidents was very low. The study also concluded, however, that the consequences of a zirconium cladding fire in a spent fuel pool could be serious and, that once the fuel was uncovered, it might take only a few hours for the most recently discharged spent fuel rods to ignite.

A paper by Alvarez et al. (2003a; see also Thompson, 2003) took the analyses in NUREG-1738 to their logical ends in light of the September 11, 2001, terrorist attacks: Namely, what would happen if there were a loss-of-pool-coolant event that drained the spent fuel pool? Such an event was not considered in NUREG-1738, but the analytical results in that study were presented in a manner that made such an analysis possible.

Alvarez and his co-authors concluded that such an event would lead to the rapid heat-up of spent fuel in a dense-packed pool to temperatures at which the zirconium alloy cladding would catch fire and release many of the fuel's fission products, particularly cesium-137. They suggested that the fire could spread to the older spent fuel, resulting in long-term contamination consequences that were worse than those from the Chernobyl accident. Citing two reports by Brookhaven National Laboratory (BNL, 1987, 1997), they estimated that between 10 and 100 percent of the cesium-137 could be mobilized in the plume from the burning spent fuel pool, which could cause tens of thousands of excess cancer deaths, loss of tens of thousands of square kilometers of land, and economic losses in the hundreds of billions of dollars. The excess cancer estimates were revised downward to between 2000 and 6000 cancer deaths in a subsequent paper (Beyea et al., 2004) that more accurately accounted for average population densities around U.S. power plants.

Alvarez and his co-authors recommended that spent fuel be transferred to dry storage within five years of discharge from the reactor. They noted that this would reduce the radioactive inventories in spent fuel pools and allow the remaining fuel to be returned to open-rack storage to allow for more effective coolant circulation, should a loss-of-pool-coolant event occur. The authors also discussed other compensatory measures that could be taken to reduce the consequences of such events.

The Alvarez et al. (2003a) paper received extensive attention and comments, including a comment from the Nuclear Regulatory Commission staff (USNRC, 2003a; see Alvarez et al., 2003b, for a response). None of the commentators challenged the main conclusion of the Alvarez et al. (2003a) paper that a severe loss-of-pool-coolant accident might lead to a spent fuel fire in a dense-packed pool. Rather, the commentators challenged the likelihood that such an event could occur through accident or sabotage, the assumptions used to calculate the offsite consequences of such an event, and the cost-effectiveness of the authors' proposal to move spent fuel into dry cask storage. One commentator summarized these differences in a single sentence (Benjamin, 2003, p. 53): "In a nutshell, [Alvarez et al.] correctly identify a problem that needs to be addressed, but they do not adequately demonstrate that the proposed solution is cost-effective or that it is optimal."

The Nuclear Regulatory Commission staff provided a briefing to the committee that provides a further critique of the Alvarez et al. (2003a) analysis that goes beyond the USNRC (2003a) paper. Commission staff told the committee that the NUREG-1738 analyses attempted to provide a bounding analysis of current and conceivable future spent fuel pools at plants undergoing decommissioning and therefore relied on conservative assumptions. The analysis assumed, for example, that the pool contained an equivalent of three-and-one-half reactor cores of spent fuel, including the core from the most recent reactor cycle. The staff also asserted that NUREG-1738 did not provide a realistic analysis of consequences. Commission staff concluded that "the risks and potential societal cost of [a] terrorist attack on spent fuel pools do not justify the complex and costly measures proposed in Alvarez et al. (2003) to move and store 1/3 of spent fuel pools [sic] inventory in dry storage casks."[8]

The committee provides a discussion of the Alvarez et al. (2003a) analysis in its classified report. The committee judges that some of their release estimates should not be dismissed.

The 2003 Nuclear Regulatory Commission (USNRC, 2003b) staff publication NUREG-0933, *A Prioritization of Generic Safety Issues*,[9] discusses beyond-design-basis accidents in spent fuel pools. The study draws some of the same consequence conclusions as the Alvarez et al. (2003a) paper. It notes that in a dense-packed pool, a zirconium cladding fire "would probably spread to most or all of the spent fuel pool" (p. 1). This could drive what the report refers to as "borderline aged fuel" into a molten condition leading to the release of fission products comparable to molten fuel in a reactor core.

The NUREG-0933 report (USNRC, 2003b) summarizes technical analyses of the frequencies of severe accidents for three BWR scenarios. The report concludes that the greatest risk is from a beyond-design-basis seismic event. While the consequences of such accidents are considerable, the report concludes that their frequencies are no greater than would be expected for reactor core damage accidents due to seismic events beyond the design basis safe shutdown earthquake.

An analysis of spent fuel operating experience by the Nuclear Regulatory Commission staff (USNRC, 1997) showed that several accidental partial-loss-of-pool-coolant events have occurred as a result of human error. Two of these involved the loss of more than 5 feet of water from the pool, but none had serious consequences. Nevertheless, Commission staff suggested that plant-specific analyses and corrective actions should be taken to reduce the potential for such events in the future.

It is important to recognize that with the exception of the Alvarez et al. (2003a) paper, all of the previous U.S. work reviewed by the committee has focused on safety risks, not security risks. The Nuclear Regulatory Commission analyses of spent fuel storage vulnerabilities were not completed by the time the committee finalized its information gathering for this report, but the committee did receive briefings on this work. In addition, analyses have been undertaken of external impacts on power plant structures by aircraft for the few commercial power plants that are located close enough to airports to consider hardening of the plant design to resist accidental aircraft crashes. These analyses were done as part of the plants' licensing safety analyses. The committee did not look further into these few plants because the aircraft considered were smaller and the impact velocities considered were much lower than those that might be brought to bear in a well-planned terrorist attack.

The committee did learn about work to assess the risks of spent fuel storage to terrorist attacks in Germany (see Appendix C for a description). However, the details of this work are classified by the German government and therefore are unavailable to the

---

[8] The quote is from a PowerPoint presentation made by Nuclear Regulatory Commission staff to the committee at one of its meetings.
[9] NUREG-0933 is a historical record that provides a yearly update of generic safety issues. It does not provide any additional technical analysis of these issues.

committee for review. Consequently, the committee was unable to provide a technical assessment.

3.3 EVALUATION OF THE POTENTIAL RISKS OF POOL STORAGE

Prior to the September 11, 2001, terrorist attacks, spent fuel pool analyses by the Nuclear Regulatory Commission were focused almost exclusively on safety. On the basis of these analyses, the Commission concluded that spent fuel storage carried risks that were no greater (and likely much lower) than risks for operating nuclear reactors, as discussed in the previous section of this chapter.

The September 11, 2001, terrorist attacks raised the possibility of a new kind of threat to commercial power plants and spent fuel storage: premeditated, carefully planned, high-impact attacks by terrorists to damage these facilities for the purpose of releasing radiation to the environment and spreading fear and panic among civilian populations. The Commission informed the committee that its conclusions about risks of spent fuel storage are now being reevaluated in light of these new threats.

Prior to September 11, the Nuclear Regulatory Commission viewed the most credible sabotage event as a violent external land assault by small groups of well-trained, heavily armed individuals aided by a knowledgeable insider.[10] The Commission has long-established requirements for physical protection systems at power plants to thwart such assaults. The committee was told that these requirements have been increased since the September 11, 2001, attacks. To the committee's knowledge, there are currently no requirements in place to defend against the kinds of larger-scale, premeditated, skillful attacks that were carried out on September 11, 2001, whether or not a commercial aircraft is involved. Staff from the Nuclear Regulatory Commission and representatives from the nuclear industry repeatedly told the committee that they view detecting, preventing, and thwarting such attacks as the federal government's responsibility.

It is important to recognize that nuclear power plants in the United States and most of the rest of the world[11] were designed primarily with safety, not security, in mind.[12] The reinforced concrete containment buildings that house the reactors were designed to contain internal pressures of up to about 4 atmospheres in case steam is released in the event of various hypothetical reactor accidents. These and other plant structures were not specifically designed to resist external terrorist attacks, although their robust construction would certainly provide significant protection against external assaults with airplanes or other types of weapons. Moreover, commercial power plants are substantially more robust than other critical infrastructure such as chemical plants, refineries, and fossil-fuel-fired electrical generating stations.

---

[10] This is known as the "design basis threat" for radiological sabotage of nuclear power plants. See Chapter 2.

[11] Spent fuel storage facilities in Germany are designed to survive the impact of a Phantom military jet without a significant release of radiation. Since September 11, 2001, the Germans have also examined the impact of a range of aircraft, including large civilian airliners, on these facilities. A discussion is provided in Appendix C.

[12] No nuclear power plant ordered after the mid-1970s has been built in the United States, so the designs were developed long before domestic terrorism of the kind seen on September 11, 2001, became a concern.

In the wake of the September 11, 2001, attacks, a great deal of additional work has been or is being carried out by government and private entities to assess the security risks posed by terrorist attacks against nuclear power plants and spent fuel storage. The committee provides a discussion of these studies in the following subsections. Some of these studies are still in progress.

The committee's discussion of this work in the following subsections is organized around the following two questions:

(1) Could an accident or terrorist attack lead to a loss-of-pool-coolant event that would partially or completely drain a spent fuel pool?
(2) What would be the radioactive releases if a pool were drained?

3.3.1 Could a Terrorist Attack Lead to a Loss-of-Pool-Coolant Event?

A terrorist attack that either disrupted the cooling system for the spent fuel pool or damaged or collapsed the pool itself could potentially lead to a loss-of-pool-coolant event. The cooling system could be disrupted by disabling or damaging the system that circulates water from the pool to heat exchangers to remove decay heat. This system would not likely be a primary target of a terrorist attack, but it could be damaged as the result of an attack on the spent fuel pool or other targets at the plant (e.g., the power for the pumps could be interrupted). The loss of cooling capacity would be of much greater concern were it to occur during or shortly after a reactor offloading operation, because the pool would contain a large amount of high decay-heat fuel.

The consequences of a damaged cooling system would be quite predictable: The temperature of the pool water would rise until the pool began to boil. Steam produced by boiling would carry away heat, and the steam would cool as it expanded into the open space above the pool.[13] Boiling would slowly consume the water in the pool, and if no additional water were added the pool level would drop. It would likely take several days of continuous boiling to uncover the fuel. Unless physical access to the pool were completely restricted (e.g., by high radiation fields or debris), there would likely be sufficient time to bring in auxiliary water supplies to keep the water level in the pool at safe levels until the cooling system could be repaired. This conclusion presumes, of course, that technical means, trained workers, and a sufficient water supply were available to implement such measures. The Nuclear Regulatory Commission requires that alternative sources of water be identified and available as an element of each plant's operating license.

The pool-boiling event described above could result in the release of small amounts of radionuclides that are normally present in pool water.[14] These radionuclides would likely have little or no offsite impacts given their small concentrations in the steam and their subsequent dilution in air once released to the environment. Moreover, as long as the spent fuel is covered with a steam-water mixture, it would not heat up sufficiently for the cladding to ignite.

A loss-of-pool-coolant event resulting from damage or collapse of the pool could

---

[13] The building above the spent fuel pool contains blow-out panels that could be removed to provide additional ventilation.
[14] This contamination may enter the water from damaged fuel or from neutron-activated materials that build up on the external surfaces of the fuel assemblies. The latter material is referred to as "crud."

have more severe consequences. Severe damage of the pool wall could potentially result from several types of terrorist attacks, for instance:

(1) Attacks with large civilian aircraft.
(2) Attacks with high-energy weapons.
(3) Attacks with explosive charges.

The committee reviewed two independent analyses of aircraft impacts on power plant structures: A study sponsored by EPRI completed in 2002 provides a generic analysis of civilian airliner impacts on commercial power plant structures (EPRI, 2002). A study in progress by Sandia National Laboratories for the Nuclear Regulatory Commission examines the consequences of an aircraft impact on an actual BWR power plant.

The EPRI and Sandia analyses used different finite element and finite difference codes that are in common use in research and industry.[15] Both sets of analyses attempted to validate the codes against physical tests, such as the Sandia "slug tests" that impacted water barrels into a concrete test wall at high speeds. EPRI's analysis used a Riera impact loading condition, which models the aircraft impact on a rigid structure and is a slightly conservative assumption because the structures are in fact deformable. The Sandia analysis was carried out on powerful computers that allowed the aircraft to be included explicitly in the calculations.

The committee also reviewed the preliminary results of Nuclear Regulatory Commission studies on the response of thick reinforced concrete walls such as those used in spent fuel pools to attacks involving simple explosive charges and other high-energy devices. The details of the analyses were not provided and therefore could not be evaluated quantitatively. However, some of these preliminary results are described in the committee's classified report.

The results of these aircraft and assault studies are classified or safeguards information. The committee has concluded that there are some scenarios that could lead to the partial failure of the spent fuel pool wall, thereby resulting in the partial or complete loss of pool coolant. A zirconium cladding fire could result if timely mitigative actions to cool the fuel were not taken. Details are provided in the classified report.

3.3.2 What would be the Radioactive Releases if a Pool Were Drained?

There are two ways in which an attack on a spent fuel pool could spread radioactive contamination: mechanical dispersion and zirconium cladding fires. An explosion or high-energy impact directly on the spent fuel could mechanically pulverize and loft fuel out of the pool. This would contaminate the plant and surrounding site with pieces of spent fuel. Large-

[15] The EPRI analyses used several finite element models (ABAQUS, LS DYNA, ANACAP, and WINFRITH) and Riera impact functions. The Sandia analyses used the CTH finite difference model and the Pronto3D finite element analysis model. The CTH code has been used for a wide range of impact penetration and explosive detonation problems by the Department of Energy, the Department of Defense, and industry during the past decade. CTH results have been compared extensively with experimental results. As an Eulerian code (where material flows through a fixed grid) it can readily handle severe distortions. It also has a variety of computational material models for dynamic (high-strain-rate) conditions, although it is limited in that it does not explicitly model structural members, such as rebar and metal liners in the concrete structure, because of computational requirements.

scale offsite releases of the radioactive constituents would not occur, however, unless they were mobilized by a zirconium cladding fire that melted the fuel pellets and released some of their radionuclide inventory. Such fires would create thermal plumes that could potentially transport radioactive aerosols hundreds of miles downwind under appropriate atmospheric conditions.

The Nuclear Regulatory Commission is now sponsoring work at Sandia National Laboratories to improve upon the analyses in NUREG-1738 (USNRC, 2001a), and in particular to obtain an improved phenomenological understanding of the thermal and hydraulic processes that would occur in a spent fuel pool from a loss-of-pool-coolant event. The committee received briefings on this work from Commission and Sandia staff during the course of this study. Additionally, the committee received a briefing from ENTERGY Corp. staff and its consultants under contract to analyze and understand the consequences of a loss-of-pool-coolant event in a spent fuel pool in a PWR plant.

The Sandia analyses were carried out on the reference BWR described in Section 3.1. Sandia's analysis of a PWR spent fuel pool had only just begun by the end of May 2004 and has not yet yielded any results. The committee had less opportunity to examine ENTERGY's approach and results. Because of these limitations, the committee was unable to examine in any detail the effects of the differences between BWR and PWR pools and fuel, except as noted with respect to their locations relative to grade.

The analyses were carried out using several well-established computer codes. The MELCOR code, which was developed by Sandia for use in analyzing severe reactor core accidents, was used to model fluid flow, heat transfer, fuel cladding oxidation kinetics, and fission product release phenomena associated with spent fuel assemblies. This code has been benchmarked against data from experiments (e.g., the FPT experiments on the Phébus test facility, and the VERCORS, CORA, and ORNL VI experiments)[16] that involve zirconium oxidation kinetics and fission product release. However, none of the experiments was designed to simulate the physical conditions in a spent fuel pool. Many of the phenomena are not significantly different in a reactor core and in a spent fuel pool, but a few important differences, particularly concerning fire propagation from hotter fuel assemblies to cooler fuel assemblies and nuclear fuel volatilities, warrant more detailed analyses or further experiments. In principle, MELCOR can perform "best-estimate" calculations that address a range of accident evolutions, accounting for temperature, availability of oxidizing air and steam,[17] and speciation and transport of radionuclides.

Sandia calculated the decay heat in the assemblies using the ANSI/ANS 5.1 code based on actual characteristics of the spent fuel (i.e., actual fuel ages, burn-ups, and locations) in the reference BWR pool. Flow and mixing behavior in the pool and reactor building enclosing the pool were modeled using a separate computational fluid dynamics (CFD) code.

Two types of analyses were carried out. A "separate effects" analysis was undertaken to examine the thermal responses of a spent fuel assembly (FIGURE 3.4) in a

[16] These experiments were designed to examine phenomena that occur in reactor cores during severe accidents. The phenomena include core degradation.
[17] Oxygen feeds the zirconium reaction and enhances release and transport of ruthenium-106, and the steam reaction releases hydrogen; whereas limited availability of oxygen starves the reaction. Steam can also entrain released fission products.

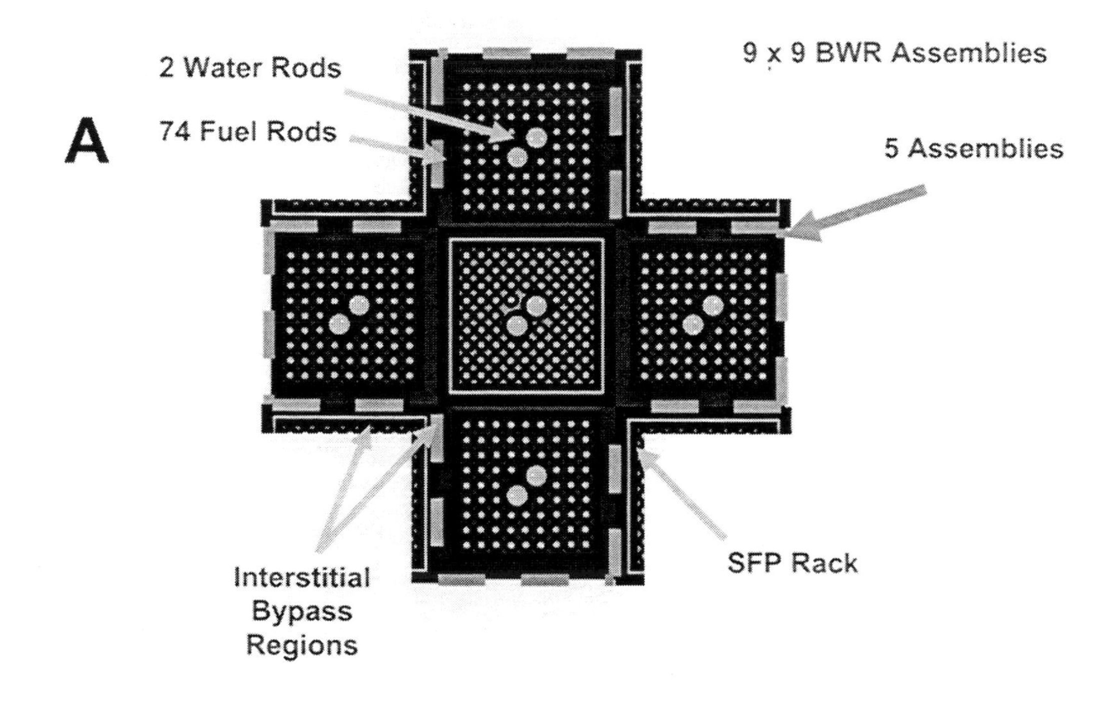

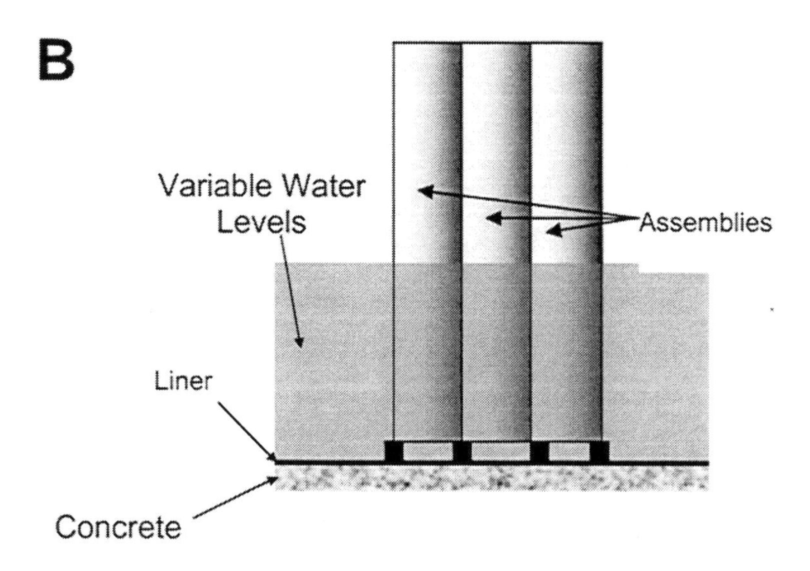

FIGURE 3.4 Configuration of fuel assemblies used for separate effects analysis. (A) Top view of BWR spent fuel assemblies used in the model. (B) Side view showing spent fuel assemblies in the pool. SOURCE: Nuclear Regulatory Commission briefing materials (2004).

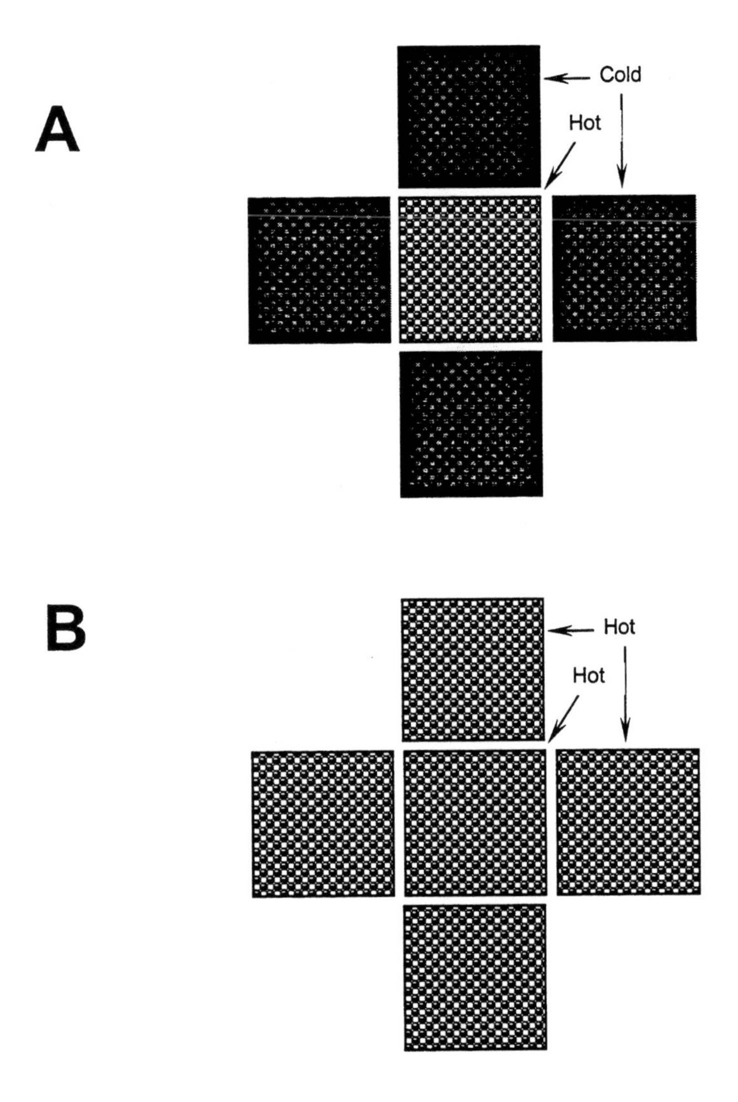

FIGURE 3.5 Two configurations used in the separate effects models shown in FIGURE 3.4: (A) Center hot spent fuel assembly surrounded by four cold assemblies; and (B) center hot spent fuel assembly surrounded by four hot assemblies. SOURCE: Nuclear Regulatory Commission briefing materials (2004).

loss-of-pool-coolant event. This analysis was used to understand how thermal behavior is influenced by factors such as decay heat in the fuel assembly, heat transfer with adjacent assemblies, and heat transfer to circulating air or steam in a drained spent fuel pool. This analysis was used to guide the development of "global response" models to examine the thermal-hydraulic behavior of an entire spent fuel pool.

The separate effects analysis examined the thermal behavior of a high decay-heat BWR spent fuel assembly surrounded either by four low decay-heat assemblies (FIGURE 3.5A) or four high decay-heat assemblies (FIGURE 3.5B). This analysis showed that the potential for heat build-up in a fuel assembly sufficient to initiate a zirconium cladding fire depends on its decay heat (which is related to its age) and on the rate at which heat can be transferred to adjacent assemblies and to circulating air or steam.

In the configuration shown in FIGURE 3.5A, the low decay-heat assemblies act as thermal radiation heat sinks, thereby allowing the more rapid transfer of heat away from the center fuel assembly than would be the case if the center assembly were surrounded by high decay-heat assemblies. The results from this analysis indicate that this configuration can be air cooled sufficiently to prevent the initiation of a zirconium cladding fire within a relatively short time after the center fuel assembly is discharged from the reactor. In the configuration shown in FIGURE 3.5B, heat transfer away from the center assembly is reduced and heat build-up is more rapid. Results indicate that this configuration cannot be air cooled for a significantly longer time after the center fuel assembly is discharged from the reactor.

The global analysis modeled the actual design and fuel loading pattern of the reference BWR spent fuel pool. The pool was divided into seven regions based on fuel age. Within each of those seven regions, the model for the fuel racks was subdivided into 16 zones. The grouping of assemblies into zones reduced the computational requirements compared to modeling every assembly.[18] Two scenarios were examined: (1) a complete loss-of-pool-coolant scenario in which the pool is drained to a level below the bottom of spent fuel assemblies; and (2) a partial-loss-of-pool-coolant scenario in which water levels in the pool drain to a level somewhere between the top and bottom of the fuel assemblies. In the former case, a convective air circulation path can be established along the entire length of the fuel assemblies, which promotes convective air cooling of the fuel. In the latter case, an effective air circulation path cannot form because the bottom of the assembly is blocked by water. Steam is generated by boiling of the pool water, and the zirconium cladding oxidation reaction produces hydrogen gas. This analysis suggests that circulation blockage has a significant impact on thermal behavior of the fuel assemblies. The specific impact depends on the depth to which the pool is drained.

The global analysis examined the thermal behavior of fuel assemblies in the pool at 1, 3, and 12 months after the offloading of one-third of a core of spent fuel from the reactor. Sensitivity studies were carried out to assess the importance of radiation heat transfer between different regions of the pool, the effects of building damage on releases of radioactive material to the environment, and the effects of varying the assumed location and size of the hole in the pool wall.

The results of these analyses are provided in the committee's classified report. For some scenarios, the fuel could be air cooled within a relatively short time after its removal from the reactor. If a loss-of-coolant event took place before the fuel could be air cooled, however, a zirconium cladding fire could be initiated if no mitigative actions were taken. Such fires could release some of the fuel's radioactive material inventory to the environment in the form of aerosols.

For a partial-loss-of-pool-coolant event, the analysis indicates that the potential for zirconium cladding fires would exist for an even greater time (compared to the complete-loss-of-pool-coolant event) after the spent fuel was discharged from the reactor because air circulation can be blocked by water at the bottom of the pool. Thermal coupling between adjacent assemblies will be due primarily to radiative rather than convective heat transfer. However, this heat transfer mode has been modeled simplistically in the MELCOR runs

---

[18] The global-response model runs took between 10 and 12 days on the personal computers used in the Sandia analyses.

performed by Sandia.[19]

If the water level is above the top of the fuel racks, decay heat in the fuel could cause the pool water to boil. Once water levels fall below a certain level in the fuel assembly, the exposed portion of the fuel cladding might heat up sufficiently to ignite if no mitigative actions were taken. This could result in the release of a substantial fraction of the cesium inventory to the environment in the form of aerosols.

A zirconium cladding fire in the presence of steam could generate hydrogen gas over the course of the event. The generation and transport of hydrogen gas in air was modeled in the Sandia calculations as was the deflagration of a hydrogen-air mixture in the closed building space above the spent fuel pool. The deflagration of hydrogen could enhance the release of radioactive material in some scenarios.

Sandia was just beginning to carry out a similar set of analyses for a "reference" PWR spent fuel pool when the committee completed information gathering for its classified report. There are reasons to believe that the results for a PWR pool could be somewhat different, and possibly more severe, than for a BWR pool: PWR assemblies are larger, have somewhat higher burn-ups, and some assemblies sit directly over the rack feet, which may impede cooling. While PWR fuel assemblies hold more fuel, they also have more open channels within them for water circulation. The committee was told that as part of this work, a sensitivity analysis will be carried out to understand how design differences among U.S. PWRs will influence the model results.

ENTERGY Corp. has carried out independent separate-effects modeling of a PWR spent fuel pool using the MELCOR code. The analyses addressed both partial and complete loss-of-pool-coolant events for its PWR spent fuel assemblies in a region of the pool where there are no water channels in the spent fuel racks. The analyses were made for relatively fresh spent fuel assemblies (i.e., separate models were run for assemblies that had been discharged from the reactor for 4, 30, and 90 days) surrounded by four "cold" assemblies that had been discharged for two years. In general, the ENTERGY results are similar to those from the Sandia separate-effects analyses mentioned above.

Several steps could be taken to mitigate the effects of such loss-of-pool-coolant events short of removal of spent fuel from the pool. Among these are the following:

- The spent fuel assemblies in the pools can be reconfigured in a "checkerboard" pattern so that newer, higher decay-heat fuel elements are surrounded by older, lower decay-heat elements. The older elements will act as radiation heat sinks in the event of a coolant loss so that the fuel is air coolable within a short time of its discharge from the reactor. Alternatively, newly discharged fuel can be placed near the pool wall, which also acts as a heat sink. ENTERGY staff estimates that reconfiguring the fuel in one of its pools into a checkerboard pattern would take only about 10 hours of extra work, but would not extend a refueling outage. Reconfiguring of fuel already in the pool could be done at any time. It does not require a reactor outage.

---

[19] In a reactor core accident, heat transfer by thermal radiation is not important because all of the fuel assemblies are at approximately the same temperature. Consequently, there is no net heat transfer between them. But spent fuel pools contain assemblies of different ages, burn-ups, and decay-heat production. The hotter assemblies will radiate heat to cooler assemblies.

- If there is sufficient space in the pool, empty slots can also be arranged to promote natural air convection in a complete-loss-of-pool-coolant event. The cask loading area in some pools may serve this purpose if it is in communication with the rest of the pool.
- Preinstalled emergency water makeup systems in spent fuel pools would provide a mechanism to replace pool water in the event of a coolant loss.
- Preinstalled water spray systems above or within the pool could also be used to cool the fuel in a loss-of-pool-coolant event.[20] The committee carried out a simple aggregate calculation suggesting that a water spray of about 50 to 60 gallons (about 190 to 225 liters) per minute for the whole pool would likely be adequate to prevent a zirconium cladding fire in a loss-of-pool-coolant event. A simple, low-pressure spray distribution experiment could verify what distribution of coolant would be sufficient to cool a spent fuel pool. Such a system would have to be designed to function even if the spent fuel pool or building were severely damaged in an attack.[21]
- Limiting full-core offloads to situations when such offloads are required would reduce the decay heat load in the pool during routine refueling outages. Alternatively, delaying the offload of fuel to the pool after a reactor shutdown would reduce the decay-heat load in the pool.
- The walls of spent fuel pools could be reinforced to prevent damage that could lead to a loss-of-pool-coolant event.
- Security levels at the plant could be increased during outages that involve core offloads.

Of course, damage to the pool and high radiation fields could make it difficult to take some of these mitigative measures. Multiple redundant and diverse measures may be required so that more than one remedy is available to mitigate a loss-of-pool-coolant event, especially when access to the pool is limited by damage or high radiation fields. Cost considerations might be significant, particularly for measures such as installing hardened spray systems and lengthening refueling outages, but the committee did not examine the costs of these measures.

3.3.3 Discussion

The Sandia and ENTERGY analyses described in this chapter were still in progress when the committee completed its classified report. As noted previously, draft technical documents describing the work were not available at the time this study was being completed. Consequently, the committee's understanding of these analyses is based on briefing materials (i.e., PowerPoint slides) presented before the committee by Nuclear

---

[20] There is an extensive analytic and experimental experience base confirming that spray systems are effective in providing emergency core cooling in BWR reactor cores, which generate much more decay heat than spent fuel. Detailed experiments have shown that some minimum amount of water must be delivered on top of each assembly, and if that is provided, the assembly will be cooled adequately even if there is significant blockage of the cooling channels.

[21] ENTERGY staff mentioned the possible use of a specially equipped fire engine to provide spray cooling. The committee does not know whether this would deliver sufficient spray cooling where it is needed or would provide sufficient protection if terrorists are attempting to prevent emergency response, but the strategy is worth further examination.

Regulatory Commission and ENTERGY staff and consultants, discussions with these experts, and the committee's own expert judgment.

The committee judges that these analyses provide a start for understanding the behavior of spent fuel pools in severe environments. The analyses were carried out by qualified experts using well-known analytical methods and engineering codes to model system behaviors. Although this is a start, the analyses have important limitations.

The aircraft attack scenarios consider one type of aircraft. Heavier aircraft could be used in such attacks. These planes are in common use in passenger and/or cargo operations, and some of these planes can be chartered.

Equally limiting assumptions were made in the analyses of spent fuel pool thermal behavior: To make the analysis tractable, it was assumed that the fuel in the pool was in an undamaged condition when the loss-of-pool-coolant event occurred. This is not necessarily a valid assumption. Whether such damage would change the outcome of the analyses described in this chapter is unknown.

Simplistic modeling assumptions were made about the fuel assembly geometry (e.g., individual fuel bundles were not modeled in the global effects calculation), convective cooling flow paths and mechanisms, thermal radiation heat transfer, propagation of cladding fires to low-power bundles, and radioactivity release mechanisms. In addition, flow blockage due to fission-gas-induced clad ballooning[22] was not considered. The thermal analysis experts on the committee judge that these simplistic assumptions could produce results that are more severe (i.e., overconservative) than would be the case had more realistic assumptions been used.

More sophisticated models, which involve clad ballooning and detailed thermal-hydraulics, including radiative heat transfer, have been developed for the analysis of severe in-core accidents. These models can be evaluated using more powerful computers. MELCOR appears to have sufficient capability to evaluate more sophisticated models of the spent fuel pool and Sandia has access to large, sophisticated computers. State-of-the-art calculations of this type are needed for the analysis of spent fuel pools so that more informed regulatory decisions can be made.

The analyses also do not consider the possibility of an attack that ejects spent fuel from the pool. The ejection of freshly discharged spent fuel from the pool might lead to a zirconium cladding fire if immediate mitigative actions could not be taken. The application of such measures could be hindered by the high radiation fields around the fuel.

While the committee judges that some attacks involving aircraft would be feasible to carry out, it can provide no assessment of the probability of such attacks. Nevertheless, analyzing their consequences is useful for informing policy decisions on steps to be taken to protect these facilities from terrorist attack.

---

[22] If a fuel rod reaches relatively high temperatures, the gases inside can cause the cladding to balloon out, restricting and even blocking coolant flow through the spaces between the rods within the assembly.

3.4 FINDINGS AND RECOMMENDATIONS

Based on its review of spent fuel pool risks, the committee offers the following findings and recommendations.

FINDING 3A: Pool storage is required at all operating commercial nuclear power plants to cool newly discharged spent fuel.

Operating nuclear power plants typically discharge about one-third of a reactor core of spent fuel every 18-24 months. Additionally, the entire reactor core may be placed into the spent fuel pool (offloaded) during outage periods for refueling. The analyses of spent fuel thermal behavior described in this chapter demonstrate that freshly discharged spent fuel generates too much decay heat to be passively air cooled. The Nuclear Regulatory Commission requires that this fuel be stored in a pool that has an active heat removal system (i.e., water pumps and heat exchangers) for at least one year as a safety matter. Current design practices for approved dry storage systems require five years' minimum decay in spent fuel pools. Although spent fuel younger than five years could be stored in dry casks, the changes required for shielding and heat removal could be substantial, especially for fuel that has been discharged for less than about three years.

FINDING 3B: The committee finds that, under some conditions, a terrorist attack that partially or completely drained a spent fuel pool could lead to a propagating zirconium cladding fire and the release of large quantities of radioactive materials to the environment. Details are provided in the committee's classified report.

It is not possible to predict the precise magnitude of such releases because the computer models have not been validated for this application.

FINDING 3C: It appears to be feasible to reduce the likelihood of a zirconium cladding fire following a loss-of-pool-coolant event using readily implemented measures.

There appear to be some measures that could be taken to mitigate the risks of spent fuel zirconium cladding fires in a loss-of-pool-coolant event. The following measures appear to have particular merit.

- Reconfiguring of spent fuel in the pools (i.e., redistribution of high decay-heat assemblies so that they are surrounded by low decay-heat assemblies) to more evenly distribute decay-heat loads. The analyses described elsewhere in this chapter suggest that the potential for zirconium cladding fires can be reduced substantially by surrounding freshly discharged spent fuel assemblies with older spent fuel assemblies in "checkerboard" patterns. The analyses suggest that such arrangements might even be more effective for reducing the potential for zirconium cladding fires than removing this older spent fuel from the pools. However, these advantages have not been demonstrated unequivocally by modeling and experiments.
- Limiting the frequency of offloads of full cores into spent fuel pools, requiring longer shutdowns of the reactor before any fuel is offloaded to allow decay-heat levels to be managed, and providing enhanced security when such offloads must

be made. The offloading of the reactor core into the spent fuel pool during reactor outages substantially raises the decay-heat load of the pool and increases the risk of a zirconium cladding fire in a loss-of-pool-coolant event. Of course, any actions that increase the time a power reactor is shut down incur costs, which must be considered in cost-benefit analyses of possible actions to reduce risks.

- Development of a redundant and diverse response system to mitigate loss-of-pool-coolant events. Any mitigation system, such as a spray cooling system, must be capable of operation even when the pool is drained (which would result in high radiation fields and limit worker access to the pool) and the pool or overlying building, including equipment attached to the roof or walls, is severely damaged.

FINDING 3D: The potential vulnerabilities of spent fuel pools to terrorist attacks are plant-design specific. Therefore, specific vulnerabilities can be understood only by examining the characteristics of spent fuel storage at each plant.

As described in the classified report, there are substantial differences in the design of PWR and BWR spent fuel pools. PWR pools tend to be located near or below grade, whereas BWR pools typically are located well above grade but are protected by exterior walls and other structures. In addition, there are plant-specific differences among BWRs and PWRs that could increase or decrease the vulnerabilities of the pools to various kinds of terrorist attacks, making generic conclusions difficult.

FINDING 3E: The Nuclear Regulatory Commission and independent analysts have made progress in understanding some vulnerabilities of spent fuel pools to certain terrorist attacks and the consequences of such attacks for releases of radioactivity to the environment. However, additional work on specific issues listed in the following recommendation is needed urgently.

The analyses carried out to date for the Nuclear Regulatory Commission by Sandia National Laboratories and by other independent organizations such as EPRI and ENTERGY have provided a general understanding of spent fuel behavior in a loss-of-pool-coolant event and the vulnerability of spent fuel pools to certain terrorist attacks that could cause such events to occur. The work to date, however, has not been sufficient to adequately understand the vulnerabilities and consequences. This work has addressed a small number of plant designs that may not be representative of U.S. commercial nuclear power plants as a whole. It has considered only a limited number of threat scenarios that may underestimate the damage that can be inflicted on the pools by determined terrorists. Additional analyses are needed urgently to fill in the knowledge gaps so that well-informed policy decisions can be made.

RECOMMENDATION: The Nuclear Regulatory Commission should undertake additional best-estimate analyses to more fully understand the vulnerabilities and consequences of loss-of-pool-coolant events that could lead to a zirconium cladding fire. Based on these analyses, the Commission should take appropriate actions to address any significant vulnerabilities that are identified. The analyses of the BWR and PWR spent fuel pools should be extended to consider the consequences of loss-of-pool-coolant events that are described in the committee's classified report.

The consequence analyses should address the following questions:

- **To what extent would such attacks damage the spent fuel in the pool, and what would be the thermal consequences of such damage?**
- **Is it feasible to reconfigure the spent fuel within pools to prevent zirconium cladding fires given the actual characteristics (i.e., heat generation) of spent fuel assemblies in the pool, even if the fuel were damaged in an attack? Is there enough space in the pools at all commercial reactor sites to implement such fuel reconfiguration?**
- **In the event of a localized zirconium cladding fire, will such rearrangement prevent its spread to the rest of the pool?**
- **How much spray cooling is needed to prevent zirconium cladding fires and prevent propagation of such fires? Which of the different options for providing spray cooling are effective under attack and accident conditions?**

Sensitivity analyses should also be undertaken to account for the full range of variation in spent fuel pool designs (e.g., rack designs, capacities, spent fuel burn-ups, and ages) at U.S. commercial nuclear power plants.

RECOMMENDATION: While the work described in the previous recommendation under Finding 3E, above, is being carried out, the Nuclear Regulatory Commission should ensure that power plant operators take prompt and effective measures to reduce the consequences of loss-of-pool-coolant events in spent fuel pools that could result in propagating zirconium cladding fires. The committee judges that there are at least two such measures that should be implemented promptly:

- Reconfiguring of fuel in the pools so that high decay-heat fuel assemblies are surrounded by low decay-heat assemblies. This will more evenly distribute decay-heat loads, thus enhancing radiative heat transfer in the event of a loss of pool coolant.
- Provision for water-spray systems that would be able to cool the fuel even if the pool or overlying building were severely damaged.

Reconfiguring of fuel in the pool would be a prudent measure that could probably be implemented at all plants at little cost, time, or exposure of workers to radiation. The second measure would probably be more expensive to implement and may not be needed at all plants, particularly plants in which spent fuel pools are located below grade or are protected from external line-of-sight attacks by exterior walls and other structures.

The committee anticipates that the costs and benefits of options for implementing the second measure would be examined to help decide what requirements would be imposed. Further, the committee does not presume to anticipate the best design of such a system—whether it should be installed on the walls of a pool or deployed from a location where it is unlikely to be compromised by the same attack—but simply notes the demanding requirements such a system must meet.

4

DRY CASK STORAGE AND COMPARATIVE RISKS

This chapter addresses the second and third charges of the committee's statement of task:

- The safety and security advantages, if any, of dry cask storage[1] versus wet pool storage at reactor sites.
- Potential safety and security advantages, if any, of dry cask storage using various single-, dual-, or multi-purpose cask designs.

The second charge calls for a comparative analysis of dry cask storage versus pool storage, whereas the third charge focuses exclusively on dry casks. The committee will address the third charge first to provide the basis for the comparative analysis.

By the late 1970s, the need for alternatives to spent fuel pool storage was becoming obvious to both commercial nuclear power plant operators and the Nuclear Regulatory Commission. The U.S. government made a policy decision at that time not to support commercial reprocessing of spent nuclear fuel (see Appendix D). At the same time, efforts to open an underground repository for permanent disposal of commercial spent fuel were proving to be more difficult and time consuming than originally anticipated.[2] Commercial nuclear power plant operators had no place to ship their growing inventories of spent fuel and were running out of pool storage space.

Dry cask storage was developed to meet the need for expanded onsite storage of spent fuel at commercial nuclear power plants. The first dry cask storage facility in the United States was opened in 1986 at the Surry Nuclear Power Plant in Virginia. Such facilities are now in operation at 28 operating and decommissioned nuclear power plants. In 2000, the nuclear power industry projected that up to three or four plants per year would run out of needed storage space in their pools without additional interim storage capacity.

This chapter is organized into the following sections:

- Background on dry cask storage.
- Evaluation of potential risks of dry cask storage.
- Potential advantages of dry storage over wet storage.
- Findings and recommendations.

[1] This storage system is referred to as "dry" because the fuel is stored out of water.
[2] The Nuclear Waste Policy Act of 1982 and the Amendments Act of 1987 laid out a process for identifying a site for a geologic repository. That repository was to be opened and operating by the end of January 1998. The federal government now hopes to open a repository at Yucca Mountain, which is located in southwestern Nevada, by the end of 2010.

4.1 BACKGROUND ON DRY CASK STORAGE

The storage of spent fuel in dry casks has the same three primary objectives as pool storage (Chapter 3):

- Cool the fuel to prevent heat-up to high temperatures from radioactive decay.
- Shield workers and the public from the radiation emitted by radioactive decay in the spent fuel and provide a barrier for any releases of radioactivity.
- Prevent criticality accidents.

Dry casks are designed to achieve the first two of these objectives without the use of water or mechanical systems. Fuel cooling is passive: that is, it relies upon a combination of heat conduction through solid materials and natural convection or thermal radiation through air to move decay heat from the spent fuel into the ambient environment. Radiation shielding is provided by the cask materials: Typically, concrete, lead, and steel are used to shield gamma radiation, and polyethylene, concrete, and boron-impregnated metals or resins are used to shield neutrons. Criticality control is provided by a lattice structure, referred to as a *basket*, which holds the spent fuel assemblies within individual compartments in the cask (FIGURE 4.1). These maintain the fuel in a fixed geometry, and the basket may contain boron-doped metals to absorb neutrons.[3]

Passive cooling and radiation shielding are possible because these casks are designed to store only older spent fuel. This fuel has much lower decay heat than freshly discharged spent fuel as well as smaller inventories of radionuclides.

The industry sometimes refers to these casks using the following terms:

- Single-, dual-, and multi-purpose casks.
- Bare-fuel and canister-based casks.

The terms in the first bullet indicate the application for which the casks are intended to be used. Single-purpose cask systems are licensed[4] only to store spent fuel. Dual-purpose casks are licensed for both storage and transportation. Multi-purpose casks are intended for storage, transportation, and disposal in a geologic repository. No true multi-purpose casks exist in the United States (or in any other country for that matter) because specifications for acceptable containers for geologic disposal have yet to be finalized by the Department of Energy. Current plans for Yucca Mountain do not contemplate the use of multi-purpose casks.

Nevertheless, some cask vendors still refer to their casks as "multi-purpose." These are at best dual-purpose casks, however, because they have been licensed only for storage and transport. **Because true multi-purpose casks do not now exist and are not likely to exist in the future, the committee did not consider them further in this study.**

---

[3] Criticality control is less of an issue in dry casks because there is no water moderator present after the cask is sealed and drained.

[4] Authority for licensing dry cask storage rests with the Nuclear Regulatory Commission.

FIGURE 4.1 Photo of NUHOMS canister showing the internal basket for holding the spent fuel assemblies in a fixed geometry. This canister is shown for illustrative purposes only. SOURCE: Courtesy of Transnuclear, Inc., an Areva Company.

The terms in the second bullet indicate how spent fuel is loaded into the casks. In bare-fuel[5] casks, spent fuel assemblies are placed directly into a basket that is integrated into the cask itself (see FIGURE 4.3B). The cask has a bolted lid closure for sealing. In canister-based casks, spent fuel assemblies are loaded into baskets integrated into a thin-wall (typically ½-inch [1.3-centimeter] thick) steel cylinder, referred to as a *canister* (see FIGURE 4.1 and 4.3A). The canister is sealed with a welded lid. The canister can be stored or transported if it is placed within a suitable overpack. This overpack is closed with a bolted lid.

Bare-fuel and canister-based systems are sometimes referred to as "thick-walled" and "thin-walled" casks, respectively, by some cask vendors. This designation is not strictly correct because the overpacks in canister-based systems have thick walls. The only thin-walled component is the canister, which is designed to be stored or transported within the overpack.

The designation of a cask as single- or dual-purpose often has less to do with its design and more to do with licensing decisions. Indeed, bare-fuel and canister-based casks can be licensed for either single or dual purposes. Consequently, one should not expect the performance of a cask in accidents or terrorist attacks to depend on its designation as single- or dual-purpose. Rather, performance will depend on the type of attack and construction of the cask. For the purposes of discussion in this chapter, therefore, the committee uses the designations "bare-fuel" and "canister-based," rather than single- or dual-purpose, when referring to various cask designs.

All bare-fuel casks in use in the United States are designed to be stored vertically. Most canister-based systems also are designed for vertical storage, but one overpack

---

[5] The term *bare fuel* refers to the entire fuel assembly, including the uranium pellets within the fuel rods.

system is designed as a horizontal concrete module (FIGURE 4.2).[6] The principal characteristics of dry cask storage systems are summarized in TABLE 4.1, which is located at the end of this chapter.

Dry casks are designed to hold up to about 10 to 15 metric tons of spent fuel. This is equivalent to about 32 pressurized water nuclear reactor (PWR) spent fuel assemblies or 68 boiling water nuclear reactor (BWR) spent fuel assemblies. Although the dimensions vary among manufacturers, fuel types (i.e., BWR or PWR fuel), and amounts of fuel stored, the casks are typically about 19 feet (6 meters) in height, 8 feet (2.5 meters) in diameter, and weigh 100 tons or more when loaded.

The casks (for bare-fuel designs) or canisters (for canister-based designs) are placed directly into the spent fuel pool for loading. After they are loaded, the canisters or casks are drained, vacuum dried, and filled with an inert gas (typically helium). The loaded canisters or casks are then removed from the pool, their outer surfaces are decontaminated,[7] and they are moved to the dry storage facility on the property of the reactor site. Loading of a single cask or canister can take up to one week. The vacuum drying process is the longest step in the loading process.

In the United States, dry casks are stored on open concrete pads within a protected area of the plant site.[8,9] This protected area may be contiguous with the protected area of the plant itself or may be located some distance away in its own protected area (see FIGURE 2.1).

According to the information provided to the committee by cask vendors, nuclear power plant operators are currently purchasing mostly dual-purpose casks for spent fuel storage. The horizontal NUHOMS cask design is one of the most-ordered designs at present (TABLE 4.3). The vendors informed the committee that cost is the chief consideration for their customers when making purchasing decisions. Cost considerations are driving the cask industry away from all-metal cask designs and toward concrete designs for storage.

---

[6] In addition, there is one modular concrete vault design in the United States: the Fort St. Vrain, Colorado, Independent Spent Fuel Storage Installation, which stores spent fuel from a high-temperature gas-cooled reactor. This reactor operated until 1989 and is now decommissioned. Because this is a one-of-a-kind facility, and the time available to the committee was short, it was not examined in this study.

[7] Small amounts of radioactive contamination are present in the cooling water in the spent fuel pool. Some of this contamination is transferred to the cask or canister surfaces when it is immersed in the pool for loading.

[8] There may be exceptions in the future. Private Fuel Storage has requested a license from the Nuclear Regulatory Commission to construct a dry cask storage facility in Utah that will store fuel from multiple reactor sites. An underground dry cask storage facility has been proposed at the Humbolt Bay power plant in California to store old, low decay-heat fuel. The underground design is being proposed primarily because the site has very demanding seismic design requirements and is possible only because the fuel to be stored generates little heat.

[9] In Germany, dry casks are stored in reinforced concrete buildings. These buildings were originally designed to provide additional radiation shielding (beyond what is provided by the cask itself) to reduce doses at plant site boundaries to background levels. Some of these buildings are sufficiently robust to provide protection against crashes of large aircraft. A subgroup of the committee visited spent fuel storage sites at Ahaus and Lingen during this study. See Appendix C for details.

FIGURE 4.2 Photo showing a canister being loaded into a NUHOMS horizontal storage module. SOURCE: Courtesy of Transnuclear, Inc., an Areva Company.

4.2 EVALUATION OF POTENTIAL RISKS OF DRY CASK STORAGE

Dry casks were designed to ensure safe storage of spent fuel,[10] not to resist terrorist attacks. The regulations for these storage systems, which are given in Title 10, Part 72 of the Code of Federal Regulations (i.e., 10 CFR 72), are designed to ensure adequate passive heat removal and radiation shielding during normal operations, off-normal events, and accidents. The latter include, for example, accidental drops or tip-overs during routine cask movements. The robust construction of these casks provides some passive protection against external assaults, but the casks were not explicitly designed with this factor in mind.[11]

The regulations in 10 CFR 72 require that dry cask storage facilities (formally referred to as Independent Spent Fuel Storage Installations, or ISFSIs) be located within a protected area of the plant site (see FIGURE 2.1). However, the protection requirements for these installations are lower than those for reactors and spent fuel pools. The guard force is required to carry side arms, and its main function is surveillance: to detect and assess threats and to summon reinforcements. If the ISFSI is within the protected area of the plant

[10] Dual-purpose casks also were designed for safe transport under the requirements of Title 10, Part 71 of the Code of Federal Regulations. The committee did not examine transport of spent fuel in this study.

[11] A recent study by the German organization GRS (Gesellschaft für Anlagen- und Reaktorsicherheit, MBH) examined the vulnerability of CASTOR-type casks to large-aircraft impacts.

it would come directly under the protection of plant's guard forces. The protected area is surrounded by vehicle barriers to protect against the detonation of a design basis threat vehicle bomb.[12]

A terrorist attack that breached a dry cask could *potentially* result in the release of radioactive material from the spent fuel into the environment through one or both of the following two processes: (1) mechanical dispersion of fuel particles or fragments; and (2) dispersion of radioactive aerosols (e.g., cesium-137). As described in Chapter 3, the latter process would have greater offsite radiological consequences. The committee evaluates the potential for both of these processes later in this chapter.

In the wake of the September 11, 2001, attacks, additional work has been or is being carried out by government and private entities to assess the security risks to dry casks from terrorist attacks. Sandia National Laboratories is currently analyzing the response of dry casks to a number of potential terrorist attack scenarios at the request of the Nuclear Regulatory Commission. The committee was briefed on these analyses at two of its meetings.

Sandia is analyzing the responses of three vertical cask designs and one horizontal design to a variety of terrorist attack scenarios (FIGURE 4.3). These designs are considered to be broadly representative of the dry casks currently licensed for storage in the United States by the Nuclear Regulatory Commission (see TABLE 4.1 at the end of this chapter). The committee received briefings on these studies by Nuclear Regulatory Commission and Sandia staff.

Several attack scenarios are being considered in the Sandia analyses. They include large aircraft impacts and assaults with various types and sizes of explosive charges and other energetic devices. Details on the large aircraft impact scenarios are provided in the classified report.

Most of this work is still in progress and has not yet resulted in reviewable documents. Consequently, the committee had to rely on discussions with the experts who are carrying out these studies and its own expert judgment in assessing the quality and completeness of this work.

4.2.1 Large Aircraft Impacts

Sandia analyzed the impact of an airliner traveling at high speed into the four cask designs shown in FIGURE 4.3. These analyses examined the consequences of impacts of the fuselage and the "hard" components of the aircraft (i.e., the engines and wheel struts) into individual casks and arrays of casks on a storage pad. The latter analysis examined the potential consequences of cask-to-cask interactions resulting from cask sliding or partial tip-over. The objectives of the analyses were first to determine whether the casks would fail (i.e., the containment would be breached) and, if so, to estimate the radioactive material releases and their health consequences.

---

[12] As noted in Chapter 2, the committee did not examine surveillance requirements or the placement or effectiveness of vehicle barriers and guard stations at commercial nuclear plants.

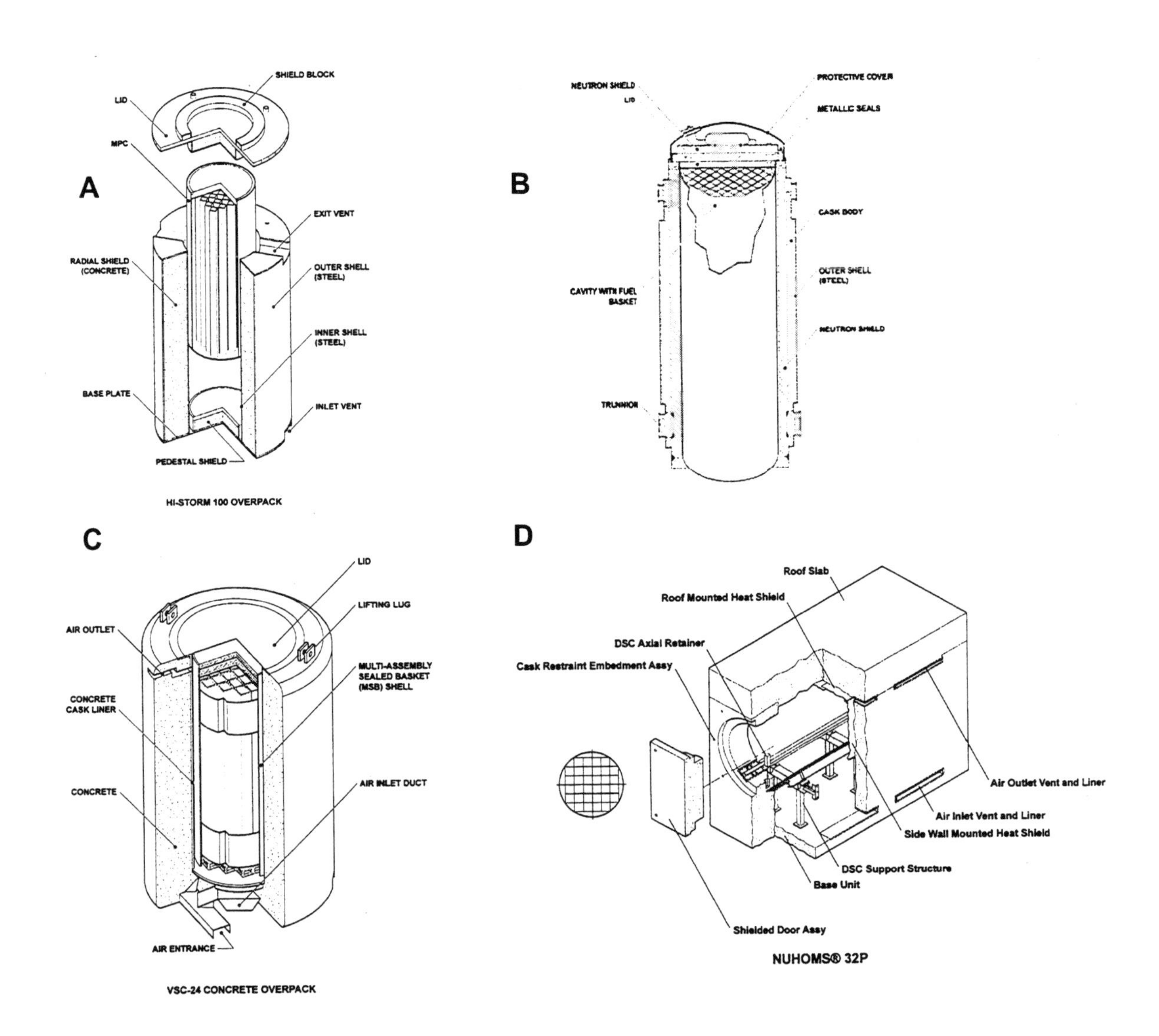

FIGURE 4.3 Four cask systems used in the Sandia analyses described in this chapter: (A) HI-STORM-100, (B) TN-68, (C) VSC-24, (D) NUHOMS-32P. The casks shown in A, C, and D are canister-based casks; the cask shown in B is a bare-fuel cask. SOURCE: Nuclear Regulatory Commission briefing materials (2004).

The aircraft was modeled using Sandia-developed Eulerian CTH code (see footnote 15 in Chapter 3). The aircraft manufacturer (Boeing Corp.) was consulted to ensure that the aircraft model used in the analyses was accurate. The casks were modeled with standard finite element codes using the published characteristics of the casks. The casks were assumed to be filled with high-burn-up, 10-year-old spent fuel. The fuel rods were assumed to fail (rupture) if the strains in the cladding exceeded 1 percent, which is a conservative assumption. Sandia evaluated the release of radioactive materials from the spent fuel pellets inside the fuel rods when such cladding failures occurred. Radiological consequences of such releases were calculated for "representative" (with respect to weather and population) site conditions for each cask based on the actual average conditions at the site that currently stores the most spent fuel in that cask type.[13] Site conditions differed for each cask.

The effects of jet fuel fires also were not considered in the analyses. Based on an analysis of actual aircraft accidents, Sandia determined that jet fuel would likely be dispersed over a large area in a low-angle impact. Consequently, the resulting petroleum fire would likely be of short duration (generally less than 15 minutes according to Sandia researchers). Long-duration fires that could damage the casks or even ignite the cladding of the spent fuel were not seen to be credible for the aircraft impact scenarios considered by Sandia.[14]

The results of these analyses, which are considered by the Nuclear Regulatory Commission to be classified or safeguards information, are detailed in the classified report. In general, the analyses show that some types of impacts will damage some types of casks. For some scenarios there could be substantial cask-to-cask interactions, including collisions and partial tip-overs.

Nevertheless, predicted releases of radioactive material from the casks, mainly noble gases, were relatively small for all of the scenarios considered by Sandia. The analyses show that the releases were governed by design-specific features of the casks. Sandia noted that the modeling of such releases is difficult and requires expert judgment for several elements of the calculation. Detailed calculations of the consequences were still in progress when the committee was briefed on these analyses.

4.2.2 Other Assaults

Analyses are also being carried out to understand the consequences of other types of assaults on the cask designs shown in FIGURE 4.3. These include assaults using explosives and certain types of high-energy devices. The analyses were still underway when the committee was briefed on these analyses, and the results were characterized by the Nuclear Regulatory Commission as preliminary. Details are provided in the classified report.

4.2.3 Discussion

As noted previously, the dry cask vulnerability analyses were still underway when the committee's classified study was completed. Based on the analyses it did receive, the committee judges that no cask provides complete protection against all types of terrorist attacks. The committee judges that releases of radioactive material from dry casks are low for the scenarios it examined with one possible exception as discussed in the classified report. It is not clear to the committee whether it is credible to assume that this "exceptional" scenario could actually be carried out.

---

[13] As noted in Chapter 1, the committee did not concern itself with how radioactive materials would be transported through the environment once they were released from a dry cask. Rather, the committee confined its examination to whether and how much radioactive material might be released from a dry cask in the event of a terrorist attack.

[14] The committee subgroup that visited Germany was briefed on a fire test on the Castor cask that involved a fully engulfing one-hour petroleum fire. The cask maintained its integrity during and after this test. See Appendix C. The results of this test do not necessarily translate to casks having other designs.

In the committee's opinion, there are several relatively simple steps that could be taken to reduce the likelihood of releases of radioactive material from dry casks in the event of a terrorist attack:

- Additional surveillance could be added to dry cask storage facilities to detect and thwart ground attacks.[15]
- Certain types of cask systems could be protected against aircraft strikes by partial earthen berms. Such berms also would deflect the blasts from vehicle bombs.
- Visual barriers could be placed around storage pads to prevent targeting of individual casks by aircraft or standoff weapons.[16] These would have to be designed so that they would not trap jet fuel in the event of an aircraft attack.
- The spacing of vertical casks on the storage pads can be changed, or spacers (shims) can be placed between the casks, to reduce the likelihood of cask-to-cask interactions in the event of an aircraft attack.
- Relatively minor changes in the design of newly manufactured casks could be made to improve their resistance to certain types of attack scenarios.

4.3 POTENTIAL ADVANTAGES OF DRY STORAGE OVER WET STORAGE

Based on the analyses presented in Chapter 3 and previously in this chapter, the committee judges that dry cask storage has several potential safety and security advantages over pool storage. These differences can best be illustrated using scenarios for both storage systems based on the Sandia analyses reviewed by the committee. **The use of such scenarios should not be taken to imply that the committee believes that these scenarios are likely or even possible at all storage facilities. They are used only for illustrative purposes.**

The following statements can be made about the comparative advantages of dry-cask storage and pool storage based on the Sandia analyses:

Less spent fuel is at risk in an accident or attack on a dry storage cask than on a spent fuel pool. An accident or attack on a dry cask storage facility would likely affect at most a few casks and put a few tens of metric tons of spent fuel at risk. An accident or attack on a spent fuel pool puts the entire inventory of the pool, potentially hundreds of metric tons of spent fuel, at risk.

The potential consequences of an accident or terrorist attack on a dry cask storage facility are lower than those for a spent fuel pool. There are several reasons for this difference:

(1) There is less fuel in a dry cask than in a spent fuel pool and therefore less radioactive material available for release.

(2) *Measured on a per-fuel-assembly basis,* the inventories of radionuclides available

---

[15] As noted in Chapter 1, the committee did not examine surveillance activities at nuclear power plants and has no basis to judge whether current activities at dry cask storage facilities are adequate.

[16] The ISFSI at the Palo Verde Nuclear Power Plant in Arizona, which was visited by a subgroup of committee members, incorporates a berm into its design to provide a visual barrier.

for release from a dry cask are lower than those from a spent fuel pool because dry casks store older, lower decay-heat fuel.

(3) Radioactive material releases from a breach in a dry cask would occur through mechanical dispersion.[17] Such releases would be relatively small. Certain types of attacks on spent fuel pools could result in a much larger dispersal of spent fuel fragments. Radioactive material releases from a spent fuel pool also could occur as the result of a zirconium cladding fire, which would produce radioactive aerosols. Such fires have the potential to release large quantities of radioactive material to the environment.

The recovery from an attack on a dry cask would be much easier than the recovery from an attack on a spent fuel pool. Breaches in dry casks could be temporarily plugged with radiation-absorbing materials until permanent fixes or replacements could be made. The most significant contamination would likely be confined largely to areas near the cask storage pad and could be detected and decontaminated. The costs of recovery could be high, however, especially if the cask could not be repaired or the spent fuel could not be removed with equipment available at the plant. A special facility might have to be constructed or brought onto the site to transfer the damaged spent fuel to other casks.

Breaches in spent fuel pools could be much harder to plug, especially if high radiation fields or the collapse of the overlying building prevented workers from reaching the pool. Complete cleanup from a zirconium cladding fire would be extraordinarily expensive, and even after cleanup was completed large areas downwind of the site might remain contaminated to levels that prevented reoccupation (see Chapter 3).

It is the potential for zirconium cladding fires in spent fuel pools that gives dry cask storage most of its comparative safety and security advantages. This comparative advantage can be reduced by lowering the potential for zirconium cladding fires in loss-of-pool-coolant events. As discussed in Chapter 3, the committee believes that there are at least two steps that can be implemented immediately to lower the potential for such fires.

4.4 FINDINGS AND RECOMMENDATIONS

With respect to the committee's task to examine potential safety and security advantages of dry cask storage using various single-, dual-, or multi-purpose cask designs, the committee offers the following findings and recommendations:

FINDING 4A: Although there are differences in the robustness of different dry cask designs (e.g., bare-fuel versus canister-based), the differences are not large when measured by the absolute magnitudes of radionuclide releases in the event of a breach.

All storage cask designs are vulnerable to some types of terrorist attacks for which radionuclide releases would be possible. The vulnerabilities are related to the specific

[17] Since the committee's classified report was published, the committee received an additional briefing from the Nuclear Regulatory Commission suggesting that a radioactive aerosol could be released in one type of terrorist attack. However, the scenario in question does not appear to the committee to be credible.

design features of the casks, but the committee judges that the quantity of radioactive material releases predicted from such attacks is still relatively small.

FINDING 4B: Additional steps can be taken to make dry casks less vulnerable to potential terrorist attacks.

Although the vulnerabilities of current cask designs are already small, additional, relatively simple steps can be taken to reduce them. Such steps are listed in Section 4.2.3.

> **RECOMMENDATION: The Nuclear Regulatory Commission should consider using the results of the vulnerability analyses for possible upgrades of requirements in 10 CFR 72 for dry casks, specifically to improve their resistance to terrorist attacks.**
>
> The committee was told by Nuclear Regulatory Commission staff that such a step is already under consideration. Based on the material presented to the committee, there appear to be minor changes that can be made by plant operators and cask vendors to increase the resistance of existing and new casks to terrorist attacks (see Section 4.2.3).

With respect to the committee's task to examine the safety and security advantages of dry cask storage versus wet pool storage at reactor sites, the committee offers the following findings and recommendations:

FINDING 4C: Dry cask storage does not eliminate the need for pool storage at operating commercial reactors.

Newly discharged fuel from the reactor must be stored in the pool for cooling, as discussed in detail in Chapter 3. Under current U.S. practices, dry cask storage can be used only to store fuel that has been out of the reactor long enough (generally greater than five years under current practices) to be air cooled. The fuel in dry cask storage poses less of a risk in the event of a terrorist attack than newly discharged fuel in pools because there is substantially reduced probability of initiating a cladding fire.

FINDING 4D: Dry cask storage for older, cooler spent fuel has two inherent advantages over pool storage: (1) It is a passive system that relies on natural air circulation for cooling; and (2) it divides the inventory of that spent fuel among a large number of discrete, robust containers. These factors make it more difficult to attack a large amount of spent fuel at one time and also reduce the consequences of such attacks.

Each storage cask holds no more than about 10 to 15 metric tons of spent fuel, compared to the several hundred metric tons of spent fuel that is commonly stored in reactor pools. The robust construction of these casks prevents large-scale releases of radionuclides in all of the attack scenarios examined by the committee. Some of the attacks could breach the casks, but many of these breaches would be small and could probably be more easily plugged than a perforated spent fuel pool wall because radiation fields would be lower and there would be no escaping water to contend with. Even large breaches of the cask would result only in the mechanical dispersal of some of its radionuclide inventory in the immediate vicinity of the cask.

FINDING 4E: Depending on the outcome of plant-specific vulnerability analyses described in the committee's classified report, the Nuclear Regulatory Commission might determine that earlier movements of spent fuel from pools into dry cask storage would be prudent to reduce the potential consequences of terrorist attacks on pools at some commercial nuclear plants.

The statement of task directs the committee to examine the risks of spent fuel storage options and alternatives for decision makers, not to recommend whether any spent fuel should be transferred from pool storage to cask storage. In fact, there may be some commercial plants that, because of pool designs or fuel loadings, may require some removal of spent fuel from their pools. If there is a need to remove spent fuel it should become clearer once the vulnerability and consequence analyses described in Chapter 3 are completed. The committee expects that cost-benefit considerations would be a part of these analyses.

TABLE 4.1 Dry Casks Used for Spent Fuel Storage in the United States

Cask design used for storage	License holder	Type	Fuel type	Construction	Closure system	Number of casks used to date; sites; and number of casks on order[1]
CASTOR V/21	GNSI (General Nuclear Systems, Inc.)	Bare-fuel, storage-only	BWR	Ductile cast iron	Primary lid (44 bolts), secondary lid (48 bolts)	25 loaded (Surry); 0 purchased
CASTOR X/33	GNS (Gesellschaft für Nuklear-Service mbH)	Bare-fuel, storage-only	PWR	Ductile cast iron	Primary lid (44 bolts), secondary lid (70 cup screws)	1 loaded (Surry); 0 purchased
NAC S/T	NAC International	Bare-fuel, storage-only	PWR	Inner and outer stainless steel shells	Closure lid (24 bolts)	2 loaded (Surry); 0 purchased
MC-10	Westinghouse	Bare-fuel, storage-only	PWR	Stainless and carbon steel	One shield lid and two sealing lids, all bolted (number of bolts not available)	1 loaded (Surry); 0 purchased
TN-32, TN-40	Transnuclear Inc.	Bare-fuel, storage-only	PWR	Carbon steel	One lid (48 bolts)	61 loaded (4 sites); 22 purchased
TN-68	Transnuclear Inc.	Bare-fuel, dual-purpose	BWR	Carbon steel	One lid (48 bolts)	24 loaded (Peach Bottom); 20 purchased
Fuel Solution W-150 Storage Cask	BNFL Fuel Solutions	Canister-based, dual-purpose	PWR, BWR	Reinforced concrete with inner steel shell	Canister lid, welded cask lid (12 bolts)	7 loaded (Big Rock Point); 0 purchased
HI-STORM 100	Holtec International	Canister-based, storage-only module	PWR, BWR	Stainless steel shells with un-reinforced concrete filler	Canister lid, welded cask lid (4 bolts)	58 loaded (7 sites); 177 on order
HI-STAR 100	Holtec International	Canister-based, dual-purpose	PWR, BWR	Carbon steel shells with neutron absorber polymer	Canister lid, welded cask lid (54 bolts)	7 loaded (2 sites[1]); 5 on order

VSC-24 Ventilated Concrete Cask	BNFL Fuel Solutions	Canister-based, storage-only	PWR	Reinforced concrete with inner steel shell	Canister lid, welded cask lid (6 bolts)	58 loaded (3 sites); 4 purchased[2]
NAC-MPC	NAC International	Canister-based, dual-purpose	PWR	Metal canister surrounded by storage overpack. Storage overpack consists of an inner steel liner 3.5 in. thick, two rebar cages, and concrete	Canister lid, welded cask lid over a shield plug (6 high-strength bolts)	21 loaded (Yankee Rowe and CT Yankee); 59 purchased
NAC-UMS	NAC International	Canister-based, dual-purpose	PWR, BWR	Metal canister surrounded by storage overpack. Storage overpack consists of inner steel liner 2.5 in. thick, two rebar cages, and concrete	Canister lid, welded cask lid over a shield plug (6 high-strength bolts)	80 loaded (2 sites); 165 purchased
Holtec MPC 24E/EF	Holtec International	Canister based, dual-purpose	PWR, BWR	Metal canister surrounded by storage overpack. Storage overpack consists of inner and outer steel liners, a double-rebar cage, and concrete	Canister lid, welded cask lid, shield plug plus 48 bolts	34 loaded (Trojan); 0 purchased
NUHOMS 24P, 52B, 61BT, 24PT1, 24PT2, 32PT	Transnuclear Inc.	Canister-based, dual-purpose	PWR, BWR	Horizontal reinforced concrete storage module with shielded canister	Canister lid, welded storage module lid, reinforced concrete	239 loaded (10 sites); >150 purchased

NOTES:
[1]The Humboldt Bay Power Plant is licensing a site-specific variation of the HI-STAR System called HI-STAR HB.
[2] Some licensees have purchased additional casks that have not yet been loaded, nor are they planned for loading.

SOURCES: Data compiled from cask license holders (2004).

5
IMPLEMENTATION ISSUES

Implementation of the recommendations in this report will require actions and cooperation by a large number of parties. This chapter provides a brief discussion of two implementation issues that the committee believes will be of interest to Congress:

(1) Timing Issues: Ensuring that high-quality, expert analyses are completed in a timely manner.
(2) Communication Issues: Ensuring that the results of the analyses are communicated to industry so that appropriate and timely mitigating actions can be taken.

5.1 TIMING ISSUES

The September 11, 2001, terrorist attacks forced the nation to begin a reexamination of the vulnerability of its critical infrastructure to high-impact suicide attacks by terrorists. The Nuclear Regulatory Commission was no exception. The Commission began a top-to-bottom review of security procedures at commercial nuclear power plants. This review resulted in the issuance of numerous directives to power plant operators to upgrade their security practices. The Commission also began a series of vulnerability analyses of spent fuel storage to terrorist attacks. These analyses are described in Chapters 3 and 4.

More than three years have passed since the September 11, 2001, attacks. Vulnerability analyses of spent fuel pool storage to attacks with large aircraft have been performed by EPRI (Chapter 3), and analyses of vulnerabilities of dry cask storage to large aircraft attacks have been completed by the German organization GRS (Gesellschaft für Anlagen- und Reaktorsicherheit, mbH). However, the Nuclear Regulatory Commission's analyses of spent fuel storage vulnerabilities have not yet been completed, and actions to reduce vulnerabilities, such as those described in Chapter 3, on the basis of these analyses have not yet been taken. Moreover, some important additional analyses remain to be done. The slow pace in completing this work is of concern given the enormous potential consequences as described elsewhere in this report.

The committee does not know the reason for this delay, nor was it asked by Congress for an evaluation. It is important to note that the Nuclear Regulatory Commission's analyses are addressing a much broader range of vulnerabilities than just spent fuel storage. The committee nevertheless raises this issue because it appears to be having an impact on the timely completion of critical work and implementation of appropriate mitigative actions for spent fuel storage.

5.2 COMMUNICATION ISSUES

During the course of this study, the committee had the opportunity to interact with representatives of the nuclear power industry to discuss their concerns about safety and security issues. The committee received numerous comments from industry representatives about the lack of information sharing by the Nuclear Regulatory Commission on the vulnerability analyses described in Chapter 3. These representatives noted that information flow was predominately in one direction: from the industry to the Commission. The Commission was not providing a reciprocal flow of information that could help the industry better understand and take early actions to address identified vulnerabilities.

Restrictions on information sharing by the Commission have resulted in missed opportunities in at least two cases observed by the committee. Analyses of aircraft impacts into power plant structures described in Chapter 3 were being carried out independently by Sandia for the Commission and by EPRI for the nuclear power industry. Because of classification restrictions, EPRI was not provided with information about the Sandia work, including the results of physical tests that would have helped EPRI validate its models. Both Sandia and the industry would have benefited had their analysts been able to talk with each other about their models, assumptions, and results while the analyses were in progress. When the EPRI work was completed the Commission declared it to be safeguards information.[1] As a consequence, some of the EPRI analysts who generated the results no longer had access to them, and the results could not be shared widely within industry.

A similar situation exists with respect to the ENTERGY Corp. spent fuel pool separate effects analyses described in Chapter 3. ENTERGY is using similar approaches and models as Sandia but has received little or no guidance from Commission staff about whether the results are realistic or consistent. The ENTERGY analysts told the committee that they would have benefited had they been able to compare and discuss their approaches and results with Sandia analysts. Sandia analysts were prevented from doing so because of classification issues. Sharing of ENTERGY's results within the company or across industry may be problematical if they are determined to be classified or safeguards information by the Commission.

Several Nuclear Regulatory Commission staff also privately expressed to the committee their frustration at the difficulty in sharing information that they know would be useful to industry. In fact, from the contacts the committee had, there does not appear to be a lack of willingness to share information at the working staff level within the Commission. Rather, it seems to be an issue of getting permission from upper management and addressing the classification restrictions.

Much of the difficulty in sharing this information appears to arise because the information is considered by the Nuclear Regulatory Commission to be safeguards information or in some cases even classified national security information. Industry analysts and decision makers generally do not have the appropriate personal security clearances[2] to access this information. The committee learned that the Commission is making efforts to share more of this information with some industry representatives. The industry will be responsible for implementing any changes to spent fuel storage to make it less vulnerable to terrorist attack. Clearly, therefore, the industry needs to understand the results of the

---

[1] Safeguards information is defined in section 147 of the Atomic Energy Act and in the Code of Federal Regulations, Title 10, Part 73.2. See the glossary for a definition. Authority for designation of safeguards resides with the Nuclear Regulatory Commission.

[2] In fact, a personnel security clearance is not required to access safeguards information. One only needs to be of "good character" and have a "need to know" as determined by the Nuclear Regulatory Commission.

Commission's vulnerability analyses to ensure that effective implementation strategies are adopted.

The committee also received complaints during this study from members of the public about the lack of information sharing. Commission staff have responded to these complaints by stating that such sharing could reveal sensitive information to terrorists and that the public does not have a "need to know" this information.

The committee fully agrees that information that could prove useful to terrorists should not be released. On the other hand, the committee believes that there is information that could be shared without compromising national security. For example, general information about the kinds of threats being considered and general steps being taken to reduce vulnerabilities could be shared with the public. Information about specific vulnerabilities of spent fuel pools and dry storage casks to terrorist attacks as well as potential mitigative actions could be shared with industry without revealing the details about how such attacks might be carried out. Sharing information with industry is essential for ensuring that mitigative actions to reduce vulnerabilities are carried out. Sharing information with the public is essential in a nation with strong democratic traditions for sustaining public confidence in the Commission as an effective regulator of the nuclear industry, and for reducing the potential for severe environmental, health, economic, and psychological consequences from terrorist attacks should they occur.

5.3 FINDING AND RECOMMENDATION

FINDING 5A: Security restrictions on sharing of information and analyses are hindering progress in addressing potential vulnerabilities of spent fuel storage to terrorist attacks.

Current classification and security practices appear to discourage information sharing between the Nuclear Regulatory Commission and industry. During the course of the study the committee received comments from power plant operators, their contractors, and Nuclear Regulatory Commission staff about the difficulties of sharing the information on the vulnerability of spent fuel storage. Indeed, even the committee found it difficult and in some cases impossible to obtain needed information (e.g., information on the design basis threat). Such restrictions have several negative consequences: They impede the review and feedback processes that can enhance the technical soundness of the analyses being carried out; they make it difficult to build support within the industry for potential mitigative measures; and they may undermine the confidence that the industry, expert panels such as this one, and the public place in the adequacy of such measures.

RECOMMENDATION: The Nuclear Regulatory Commission should improve the sharing of pertinent information on vulnerability and consequence analyses of spent fuel storage with nuclear power plant operators and dry cask storage system vendors on a timely basis.

Implementation of this recommendation will allow timely mitigation actions. Certain current security practices may have to be modified to carry out this recommendation.

The committee also believes that the public is an important audience for the work being carried out to assess and mitigate vulnerabilities of spent fuel storage facilities. While it would be inappropriate to share all information publicly, more constructive interaction with the public and independent analysts could improve the work being carried out and also increase public confidence in Nuclear Regulatory Commission and industry decisions and actions to reduce the vulnerability of spent fuel storage to terrorist threats.

REFERENCES

Alvarez, R., J. Beyea, K. Janberg, J. Kang, E. Lyman, A. Macfarlane, G. Thompson, and F. N. von Hippel. 2003a. Reducing the Hazards from Stored Spent Power-Reactor Fuel in the United States. Science and Global Security, Vol. 11, pp. 1-51.

Alvarez, R., J. Beyea, K. Janberg, J. Kang, E. Lyman, A. Macfarlane, G. Thompson, and F. N. von Hippel. 2003b. Response by the authors to the NRC review of "Reducing the Hazards from Stored Spent Power-Reactor Fuel in the United States." Science and Global Security, Vol.11, pp. 213-223.

American Nuclear Society. 1988. Design Criteria for an Independent Spent Fuel Storage Installation (Water Pool Type): An American National Standard. ANSI/ANS-57.7-1988. American Nuclear Society. LaGrange Park, Illinois.

ASCE (American Society of Civil Engineers). 2003. The Pentagon Building Performance Report. By P. F. Mlakar, D. O. Dusenberry, J. R. Harris, G. Haynes. L. T. Phan, and M. A. Sozen. January. Structural Engineering Institute. Reston, Virginia. Available at *http://fire.nist.gov/bfrlpubs/build03/art017.html.*

Baker, L., and L. C. Just. 1962. Studies of Metal Water Reactions at High Temperatures III. Experiments and Theoretical Studies of the Zirconium-Water Reaction. ANL-548. May. Argonne National Laboratory, Argonne, Illinois.

Benjamin, A. S., D. J. McCloskey, D. A. Powers, and S. A. Dupree. 1979. Spent Fuel Heatup Following Loss of Water During Storage. NUREG/CR-0649, SAND77-1371. Rev.3. Sandia National Laboratories, New Mexico.

Benjamin, A. S. 2003. Comments on "Reducing the Hazards from Stored Spent Power-Reactor Fuel in the United States." Science and Global Security, Vol. 11, pp. 53-58.

Beyea, J., E. Lyman, and F. von Hippel. 2004. Damages from a Major Release of ^{137}Cs into the Atmosphere of the U.S. (addendum to "Reducing the Hazards from Stored Spent Power-Reactor Fuel in the United States" by R. Alvarez, J. Beyea, K. Janberg. E. Lyman, A. Macfarlane, G. Thompson, and F. von Hippel, 2003. Science and Global Security, Vol. 11, pp. 1-51). Science and Global Security, Vol. 12, pp. 125-136.

Borenstein, S. 2002. Security Upgrades at Nuclear Plants Are Behind Schedule. Knight Ridder Newspapers. April 11. Available at *http://www.nci.org/02/04f/12-01.htm.*

BNL (Brookhaven National Laboratory). 1987. Severe Accidents in Spent Fuel Pools in Support of Generic Safety Issue 82. NUREG/CR-4982 and BNL-NUREG-52093. V. L. Sailer, K.R. Perkins, J.R. Weeks, and H. R. Connell. July. Upton, N.Y.: Brookhaven National Laboratory.

BNL. 1997. A Safety and Regulatory Assessment of Generic BWR and PWR Permanently Shutdown Nuclear Power Plants. R. J. Travis, R. E. Davis, E. J. Grove, and M. A. Azarm. NUREG/CR-6451. August. Upton, N.Y.: Brookhaven National Laboratory.

Chapin, D. M., K. P. Cohen, W. K. Davis, E. E. Kintner, L. J. Koch, J. W. Landis, M. Levenson, I. H. Mandil, Z. T. Pate, T. Rockwell, A. Schriesheim, J. W. Simpson, A. Squire, C. Starr, H. E. Stone, J. J. Taylor, N. E. Todreas, B. Wolfe, and E. L. Zebroski. 2002. Nuclear Power Plants and Their Fuel as Terrorist Targets. Science, Vol. 297, pp. 1997-1999.

Droste, B., H. Völzke, G. Wieser, and L. Quiao. 2002. Safety Margins of Spent Fuel Transport and Storage Casks Considering Aircraft Crash Impacts. RAMTRANS, Vol. 13(3-4), pp. 313-316.

Duderstadt, J. J., and L. J. Hamilton. 1976. Nuclear Reactor Analysis. John Wiley & Sons. New York.

EPRI. 2002. Deterring Terrorism: Aircraft Crash Impact Analyses Demonstrate Nuclear Power Plant's Structural Strength. Palo Alto, California [SAFEGUARDS INFORMATION].

FEMA (Federal Emergency Management Agency). 2002. World Trade Center Building Performance Study: Data Collection, Preliminary Observations, and Recommendations. FEMA 403. May. FEMA Region II, New York. Available at *http://www.fema.gov/library/wtcstudy.shtm*.

Ferguson, C. D., W. C. Potter, A. Sands, L. S. Spector, and F. L. Wehling. 2004. The Four Faces of Nuclear Terrorism. Center for Nonproliferation Studies. Monterey Institute of International Studies. Nuclear Threat Initiative. Monterey, California. Available at *http://cns.miis.edu/pubs/books/pdfs/4faces.pdf*.

GAO (U.S. Government Accountability Office). 2003. Spent Nuclear Fuel: Options Exist to Further Enhance Security. GAO-03-426. July. Available at *http://www.gao.gov/new.items/d03426.pdf*.

HSK (Die Hauptabteilung für die Sicherheit der Kernanlagen). 2003. Position of the Swiss Federal Nuclear Safety Inspectorate Regarding the Safety of the Swiss Nuclear Power Plants in the Event of an Intentional Aircraft Crash. HSK-AN-4626. March. Würenlingen, Switzerland.

Jenkins, B. M. 1975. Will Terrorists Go Nuclear? RAND Corporation. RAND P-5541. Santa Monica, California.

Jenkins, B. M. 1985. Will Terrorists Go Nuclear? Orbis, Vol. 29(3), pp. 507-516.

Kaplan, S., and B.J. Garrick. 1981. On the quantitative definition of risk. Risk Analysis, Vol. 1(1), pp. 11-27.

Lamarsh, J. R. 1975. Introduction to Nuclear Engineering. Addison-Wesley Publishing Company. Reading, Massachusetts.

Lange, F., G. Pretzsch, J. Dohler, E. Horman, H. Busch, and W. Koch. 1994. Experimental Determination of UO_2-Release from Spent Fuel Transport Cask after Shaped Charge Attack. INMM Annual Meeting. Naples, Florida, Vol. XXIII, pp. 408-413.

Lange, F., G. Pretzsch, E. Hörmann, and W. Koch. 2001. Experiments to Quantify Potential Releases and Consequences from Sabotage Attack on Spent Fuel Casks. Thirteenth International Symposium on Packaging and Transportation of Radioactive Materials PATRAM. Chicago, Illinois.

Lange, F., H. J. Fett, E. Hormann, E. Schrodl, G. Schwarz, B. Droste, H. Volzke, G. Wieser, and L. Qiao. 2002. Safety Margins of Transport and Storage Casks for Spent Fuel Assemblies and HAW Canisters under Extreme Accident Loads and Effects from External Events. Report within framework of Project SR 2415. April. Gesellschaft für Anlagen- und Reaktorsicherheit (GRS) mbH, Koln; Bundesanstalt für Materialforschung und -prüfung (BAM), Berlin, Germany.

Luna, R. E. 2000. Comparison of Results from Two Spent Fuel Sabotage Source Term Experiments. RAMTRANS. Vol. 11(3), pp. 261-265.

Marsh, G. E. and G. S. Stanford. 2001. National Policy Analysis #374: Terrorism and Nuclear Power: What are the Risks? National Center for Policy Research. November. Available at *http://www.nationalcenter.org/NPA374.html*.

NRC (National Research Council). 2002. Making the Nation Safer: The Role of Science and Technology in Countering Terrorism. National Academy Press. Washington, D.C.

RBR Consultants, Inc.. 2003. Terrorist Aircraft Strikes at Indian Point Spent Fuel Pools. February. Herschel Specter's testimony to the New York City Council's Committee on Environmental Protection. February. New York.

Thomauske, B. 2003. Realization of the German Concept for Interim Storage of Spent Nuclear Fuel—Current Situation and Prospects. Waste Management '03 Conference. February 23-27, 2003. Tucson, Arizona.

Thompson, G. 2003. Robust Storage of Spent Nuclear Fuel: A Neglected Issue of Homeland Security. Institute for Resource and Security Studies. Report commissioned by Citizens Awareness Network. January. Cambridge, Massachusetts.

Tong, L. S., and J. Weisman. 1996. Thermal Analysis of Pressurized Water Reactors. Third Edition. American Nuclear Society. LaGrange Park, Illinois.

U.S. Atomic Energy Commission. 1975. Reactor Safety Study. An Assessment of Accident Risks in U.S. Commercial Nuclear Power Plants. WASH-1400. August. Washington, D.C.

USNRC (U.S. Nuclear Regulatory Commission). 1976. Final Generic Environmental Statement on the Use of Recycled Plutonium in Mixed Oxide Fuel in Light-Water Cooled Reactors (GESMO). NUREG-0002. Washington, DC.

USNRC. 1983. A Prioritization of Generic Safety Issues. NUREG-0933. December. Vol. 3.82, pp. 1-6. Washington, D.C.

USNRC. 1984. Spent Fuel Heat Generation in an Independent Spent Fuel Storage Installation. Regulatory Guide 3.54 (Task CE 034-4). Office of Nuclear Regulatory Research. September. Washington, D.C.

USNRC. 1987. Case Histories of West Valley Spent Fuel Shipments. NUREG/CR-4847. January, Washington, D.C.

USNRC. 1996. Refueling Practice Survey: Final Report. May. Washington, DC. Available at *http://www.nrc.gov/reading-rm/doc-collections/news/1996/96-074.html*.

USNRC. 1997. Operating Experience Feedback Report. Assessment of Spent Fuel Cooling. NUREG-1275. Vol. 12. J. G. Ibarra, W. R. Jones, G. F. Lanik, H. L. Ornstein, S. V. Pullani. Office for Analysis and Evaluation of Operational Data. Washington, D.C.

USNRC. 2001a. Technical Study of Spent Fuel Pool Accident Risk at Decommissioning Nuclear Power Plants. NUREG-1738. Division of Systems Safety and Analysis. January. Washington, D.C.

USNRC. 2001b. Review of NRC's Dry Cask Storage Program. Audit Report. OIG-01-A-11. Office of the Inspector General. June 20. Washington, D.C.

USNRC. 2003a. Nuclear Regulatory Commission (NRC) review of "Reducing the Hazards from Stored Spent Power-Reactor Fuel in the United States." Science and Global Security, Vol. 11, pp. 203-211.

USNRC. 2003b. A Prioritization of Generic Safety Issues. NUREG-0933. R. Emrit, R. Riggs, W. Milstead, J. Pittman, and H. Vendermolen. Office of Nuclear Regulatory Research. October. Washington, DC. Available at *http://www.nrc.gov/reading-rm/doc-collections/nuregs/staff/sr0933*.

Walker, J.S. 2004. Three Mile Island: A Nuclear Crisis in Historical Perspective. University of California Press. Berkeley, California.

Zimmerman, P. D., and C. Loeb. 2004. Dirty Bombs: The Threat Revisited. Defense Horizons. Vol. 38 (January), pp. 1-11.

A

INFORMATION-GATHERING SESSIONS

The committee organized several meetings and tours to obtain information about the safety and security of spent fuel storage. A list of these meetings and tours is provided below. The committee held several *data-gathering sessions not open to the public* to obtain classified and safeguards information about the safety and security of spent fuel storage. The committee also held several *data-gathering sessions open to the public* to receive unclassified briefings from industry, independent analysts, and other interested parties including members of the public. The written materials (e.g., PowerPoint presentations and written statements) obtained by the committee at these open sessions are posted on the web site for this project: *http://dels.nas.edu/sfs*.

A.1 FIRST MEETING, FEBRUARY 12-13, 2004, WASHINGTON, D.C.

The objective of this meeting was to obtain background information on the study request from staff of the House Committee on Appropriations, Energy and Water Development Subcommittee. The committee also was briefed by one of the sponsors of the study and by two independent experts. The following is the list of topics and speakers for the open session:

- Background on the congressional request for this study. Speaker: Kevin Cook, Professional Staff, House Committee on Appropriations, Energy and Water Development Subcommittee.
- Reducing the hazard from stored spent power-reactor fuel in the United States. Speakers: Frank von Hippel, Princeton University, and Klaus Janberg, independent consultant, co-authors of the paper entitled "Reducing the Hazard from Stored Spent Power-Reactor Fuel in the United States" (Alvarez et al., 2003).
- Nuclear power plants and their fuel as terrorist targets. Speaker: Ted Rockwell, MPR Associates, Inc., co-author of the paper entitled "Nuclear Power Plants and Their Fuel as Terrorist Targets" (Chapin et al., 2002).
- Nuclear Regulatory Commission analyses of spent fuel safety and security. Speaker: Farouk Eltawila, director, Division of Systems Analysis and Regulatory Effectiveness, Office of Research, Nuclear Regulatory Commission.

On the second day of the meeting, the committee held a data-gathering session not open to the public to obtain classified briefings from the U.S. Nuclear Regulatory Commission about its ongoing analyses of spent fuel storage security.

A.2 SECOND MEETING, MARCH 4-6, 2004, ARGONNE, ILLINOIS

During the second meeting, the committee held a data-gathering session not open to the public to receive classified briefings on spent fuel storage security from the U.S. Nuclear Regulatory Commission. The committee also toured the Dresden and Braidwood Nuclear

Generating Stations to see first-hand how spent fuel is managed and stored. The two plants were chosen because of the differences in their spent fuel storage facilities.

A.3 THIRD MEETING, APRIL 15-17, 2004, ALBUQUERQUE, NEW MEXICO

During the third meeting, the committee held a data-gathering session not open to the public to receive a briefing from EPRI on spent fuel storage vulnerabilities. The committee also held a data-gathering session open to the public to receive briefings on dry cask storage systems and radioactive releases from damaged spent fuel storage casks.

- Speakers on dry cask storage systems: William McConaghy (GNB-GNSI); Steven Sisley (BNFL); Alan Hanson (Transnuclear Inc.); Charles Pennington (NAC International); and Brian Gutherman (Holtec International, via telephone).
- Radionuclide releases from damaged spent fuel. Speaker: Robert Luna, Sandia National Laboratories (retired).

A.4 TOUR OF SELECTED SPENT FUEL STORAGE INSTALLATIONS IN GERMANY

On April 25-28, 2004, a group of committee members traveled to Germany to meet with German officials and to visit selected spent fuel storage installations. The agenda of the tour was as follows:

- Meeting with Michael Sailer, chairman of the German reactors safety commission (RSK, Reaktorsicherheitskommission).
- Visit to the dry cask manufacturer GNB (Gesellschaft für Nuklear-Behälter mbH) headquarters in Essen and the cask assembly facility and test museum in Mülheim.
- Tour of the Ahaus intermediate dry storage facility.
- Meeting with Florentin Lange, GRS (Gesellschaft für Anlagen- und Reaktorsicheheit mbH), co-author of the study entitled "Safety Margins of Transport and Storage Casks for Spent Fuel Assemblies and HAW Canisters Under Extreme Accident Loads and Effects from External Events" (Lange et al., 2002).
- Tour of the Lingen nuclear power plant and its spent fuel storage facilities.

A summary of information gathered during the tour is provided in Appendix C.

A.5 FOURTH MEETING, MAY 10-12, 2004, WASHINGTON, D.C.

During the fourth meeting, the committee held a data-gathering session not open to the public to hold in-depth technical discussions with Sandia National Laboratories staff and contractors on their spent fuel storage vulnerability analyses. The committee also received an intelligence briefing from Department of Homeland Security staff on terrorist capabilities and from the U.S. Nuclear Regulatory Commission staff on terrorist scenarios.

The meeting also included a data-gathering session open to the public that included the following briefings:

- Summary of the field trip to Germany. Speaker: Louis Lanzerotti (committee chair).
- Vulnerabilities of spent nuclear fuel pools to terrorist attacks: Issues with the design basis threat. Speaker: Peter Stockton, Project on Government Oversight.
- Consequences of a major release of ^{137}Cs into the atmosphere. Speaker: Jan Beyea, Consulting in the Public Interest.

A.6 FIFTH MEETING, MAY 26-28, 2004, WASHINGTON, D.C.

The objective of this closed meeting (i.e., open only to committee members and staff) was to finalize the classified report for National Research Council review.

A.7 TOURS OF SELECTED SPENT FUEL STORAGE FACILITIES AT U.S. NUCLEAR POWER PLANTS

On June 11 and June 14, 2004, respectively, committee subgroups visited the Palo Verde Nuclear Generating Station in Arizona and the Indian Point Nuclear Generating Station in New York.

A.8 SIXTH MEETING, JUNE 28-29, 2004

The objective of this closed meeting was to complete work on the classified report.

A.9 SEVENTH MEETING, AUGUST 12-13, 2004

The objective of this closed meeting was to develop a public version of the committee's report. The committee also held a data-gathering session not open to the public to receive a briefing from the Department of Homeland Security on steps being taken to address the findings and recommendations in the classified report.

A.10 EIGHTH MEETING, OCTOBER 28-29, 2004

The objective of this closed meeting was to continue work to develop a public version of the committee's report. The committee also held a data-gathering session not open to the public to receive a briefing from the Nuclear Regulatory Commission on steps being taken to address the findings and recommendations in the classified report.

A.11 NINTH MEETING, NOVEMBER 29-30, 2004

The objective of this closed meeting was to continue work to develop a public version of the committee's report.

A.12 TENTH MEETING, January 24-25, 2005

The objective of this closed meeting was to continue work to develop a public version of the committee's report. The committee also held a data-gathering session not open to the public to meet with three commissioners from the Nuclear Regulatory Commission (Chairman Nils Diaz and members Edward McGaffigan and Jeffrey Merrifield) to discuss what additional information the commission might be willing to make available to the committee on human-factors-related issues.

REFERENCES

Alvarez, R., J. Beyea, K. Janberg, J. Kang, E. Lyman, A. Macfarlane, G. Thompson, and F. N. von Hippel. 2003a. Reducing the Hazards from Stored Spent Power-Reactor Fuel in the United States. Science and Global Security, Vol. 11, pp. 1-51

Chapin, D. M., K. P. Cohen, W. K. Davis, E. E. Kintner, L. J. Koch, J. W. Landis, M. Levenson, I. H. Mandil, Z. T. Pate, T. Rockwell, A. Schriesheim, J. W. Simpson, A. Squire, C. Starr, H. E. Stone, J. J. Taylor, N. E. Todreas, B. Wolfe, and E. L. Zebroski. 2002. Nuclear Power Plants and Their Fuel as Terrorist Targets. Science, Vol. 297, pp. 1997-1999.

Lange, F., H. J. Fett, E. Hormann, E. Schrodl, G. Schwarz, B. Droste, H. Volzke, G. Wieser, and L. Qiao. 2002. Safety Margins of Transport and Storage Casks for Spent Fuel Assemblies and HAW Canisters under Extreme Accident Loads and Effects from External Events. Report within framework of Project SR 2415. April. Gesellschaft für Anlagen- und Reaktorsicherheit (GRS) mbH, Koln; Bundesanstalt für Materialforschung und -prüfung (BAM), Berlin, Germany.

B

BIOGRAPHICAL SKETCHES OF COMMITTEE MEMBERS

LOUIS J. LANZEROTTI, *Chair*, is an expert in geophysics and electromagnetic waves and a veteran of over 40 National Research Council (NRC) studies. He currently consults for Bell Laboratories, Lucent Technologies, and is a distinguished professor for solar-terrestrial research at the New Jersey Institute of Technology. Previously, he was a distinguished member of the technical staff at Bell Labs. His research interests include space plasmas and engineering problems related to the impacts of atmospheric and space processes on telecommunications on commercial satellites and transoceanic cables. He has been associated with numerous National Aeronautics and Space Administration (NASA) space missions as well, including Voyager, Ulysses, Galileo, and Cassini, and with commercial space satellite missions to research design and operational problems associated with spacecraft and cable operations. In 1988, he was elected to the National Academy of Engineering for his work on energetic particles and electromagnetic waves in the earth's magnetosphere, including their impact on space and terrestrial communication systems. He has twice received the NASA Distinguished Public Service Medal and has a geographic feature in Antarctica named in his honor. He was appointed to the National Science Board by President George W. Bush in 2004. Dr. Lanzerotti holds a Ph.D. in physics from Harvard University.

CARL A. ALEXANDER is an expert in the behavior of nuclear material at high temperatures and also in biological and chemical weapons. He is chief scientist and senior research leader at the Battelle Memorial Institute in Columbus, Ohio. Dr. Alexander worked on fuel design and behavior for the aircraft nuclear propulsion program and several space nuclear power projects, including the Viking, Voyager, and Cassini missions. He helped analyze the evolution of the Three Mile Island accident and is involved in the French Phebus fission product experiments, which are to reproduce all of the phenomena involved during a nuclear power reactor core meltdown accident. He has served as a consultant to the Nuclear Regulatory Commission and, in the 1970s, worked on the first experiments on the effects of an attack on spent fuel shipping containers using shaped charges. He currently leads research projects on agent neutralization and collateral effects for weapons of mass destruction for the Defense Threat Reduction Agency and the Navy, and on lethality of missile defense technologies for the Missile Defense Agency. Dr. Alexander has taught materials science and engineering at the Ohio State University and has served as graduate advisor and adjunct professor at the Massachusetts Institute of Technology, University of Southampton in the United Kingdom, and the University of Maryland. He has authored over 100 peer-reviewed articles and technical reports, many of which are classified. He holds a Ph.D. in materials science from Ohio State University.

ROBERT M. BERNERO is a nuclear engineering and regulatory expert. He is now an independent consultant after retiring from the U.S. Nuclear Regulatory Commission (USNRC) in 1995. In 23 years of service for the USNRC Mr. Bernero held numerous positions in reactor licensing, fuel cycle facility licensing, engineering standards development, risk assessment research, and waste management. His final position at USNRC was as director of the Office of Nuclear Materials Safety and Safeguards. Prior to joining the USNRC he worked for the General Electric Company in nuclear technology for 13 years. He has served as a member of the Commission of Inquiry for an International

Review of Swedish Nuclear Regulatory Activities, and he currently consults on nuclear safety-related matters, particularly regarding nuclear materials licensing and radioactive waste management. Mr. Bernero received his B.A. degree from St. Mary of the Lake (Illinois), a B.S. degree from the University of Illinois, and an M.S. degree from Rensselaer Polytechnic Institute.

M. QUINN BREWSTER is an expert in energetic solids and heat transfer. He is currently the Hermia G. Soo Professor of Mechanical Engineering at the University of Illinois at Urbana-Champaign. He is involved in the Academic Strategic Alliance Program, whose objective is to develop integrated software simulation capability for coupled, system simulation of solid rocket motors including internal ballistics (multi-phase, reacting flow) and structural response (propellant grain and motor case). Dr. Brewster has authored one book on thermal radiative transfer and chapters in four other books as well as several publications on combustion science. He is a fellow of the American Society of Mechanical Engineers and associate fellow of the American Institute of Aeronautics and Astronautics. Dr. Brewster holds a Ph.D. in mechanical engineering from the University of California at Berkeley.

GREGORY R. CHOPPIN is an actinide elements and radiochemistry expert. He is currently the R.O. Lawton Distinguished Professor Emeritus of Chemistry at Florida State University. His research interests involve the chemistry and separation of the f-elements and the physical chemistry of concentrated electrolyte solutions. During a postdoctoral period at the Lawrence Radiation Laboratory, University of California, Berkeley, he participated in the discovery of mendelevium, element 101. His research and educational activities have been recognized by the American Chemical Society's Award in Nuclear Chemistry, the Southern Chemist Award of the American Chemical Society, the Manufacturing Chemist Award in Chemical Education, the Chemical Pioneer Award of the American Institute of Chemistry, a Presidential Citation Award of the American Nuclear Society, the Becquerel Medal, British Royal Society, and honorary D.Sc. degrees from Loyola University and the Chalmers University of Technology (Sweden). Dr. Choppin previously served on the NRC's Board on Chemical Sciences and Technology and Board on Radioactive Waste Management. He holds a Ph.D. in inorganic chemistry from the University of Texas, Austin.

NANCY J. COOKE is an expert in the development, application, and evaluation of methodologies to elicit and assess individual and team knowledge. She is currently a professor in the applied psychology program at Arizona State University East. She also holds a National Research Council Associateship position with Air Force Research Laboratory and serves on the board of directors of the Cognitive Engineering Research Institute in Mesa, Arizona. Her current research areas are the following: cognitive engineering, knowledge elicitation, cognitive task analysis, team cognition, team situation awareness, mental models, expertise, and human-computer interaction. Her most recent work includes the development and validation of methods to measure shared knowledge and team situation awareness and research on the impact of cross- training, distributed mission environments, and workload on team knowledge, process, and performance. This work has been applied to team cognition in unmanned aerial vehicle and emergency operation center command-and-control. She contributed to the creation of the Cognitive Engineering Research on Team Tasks Laboratory to develop, apply, and evaluate measures of team cognition. She has authored or co-authored over 70 articles, chapters, and technical reports on measuring team cognition, knowledge elicitation, and human-computer interaction. Dr. Cooke holds a Ph.D. in cognitive psychology from New Mexico State University, Las Cruces.

GORDON R. JOHNSON is an expert in penetration mechanics and computational mechanics. He is currently a senior scientist and manager of the solid mechanics group at Network Computing Services. His recent work has included the development of computational mechanics codes that include finite elements and meshless particles. He has also developed computational material models to determine the strength and failure characteristics of a variety of materials subjected to large strains, strain rates, temperatures, and pressures. His work for the U.S. Departments of Energy and Defense has included a wide range of intense impulsive loading computations for high-velocity impact and explosive detonation. He was a chief engineering fellow during his 35 years at Alliant Techsystems (formerly Honeywell). He has served as a technical advisor for university contracts with the Army Research Office, and an industry representative for its strategic planning, and was a member of the founding board of directors for the Hypervelocity Impact Society. Dr. Johnson holds a Ph.D. in structures from the University of Minnesota, Minneapolis.

ROBERT P. KENNEDY has expertise in structural dynamics and earthquake engineering. He is currently an independent consultant in structural mechanics and engineering. Dr. Kennedy has worked on static and dynamic analysis and the design of special-purpose civil and mechanical-type structures, particularly for the nuclear, petroleum, and defense industries. He has designed structures to resist extreme loadings, including seismic loadings, missile impacts, extreme winds, impulsive loads, and nuclear environmental effects, and he has developed computerized structural analysis methods. He also served as a peer reviewer for an EPRI study on aircraft impacts on nuclear power plants. In 1991, he was elected to the National Academy of Engineering for developing design procedures for civil and mechanical structures to resist seismic and other extreme loading conditions. Dr. Kennedy holds a Ph.D. in structural engineering from Stanford University.

KENNETH K. KUO is an expert in combustion, rocket propulsion, ballistics, and fluid mechanics. He is a Distinguished Professor of Mechanical Engineering at the Pennsylvania State University. He is also the leader and director of the university's High Pressure Combustion Laboratory, a laboratory with advanced instrumentation and data acquisition devices. Dr. Kuo has directed team research projects in propulsion and combustion studies for 32 years. He has edited eight books and authored one book on combustion, published over 300 technical articles, and served as principal investigator for more than 70 projects, including a Multidisciplinary University Research Initiative (MURI) grant from the U.S. Army on "Ignition and Combustion of High Energy Materials." He is now serving as principal investigator and co-principal investigator for two MURI programs on rocket and energetic materials. In 1991, he was elected fellow of American Institute of Aeronautics and Astronautics and has received several awards for his work on solid propellants combustion processes. Dr. Kuo holds a Ph.D. in aerospace and mechanical sciences from Princeton University.

RICHARD T. LAHEY, JR., is an expert in multiphase flow and heat transfer technology, nuclear reactor safety, and the use of advanced technology for industrial applications. He is currently the Edward E. Hood Professor of Engineering at Rensselaer Polytechnic Institute (RPI) and was previously chair of the Department of Nuclear Engineering and Science, director of the Center for Multiphase Research, and the dean of engineering at RPI. Previously, Dr. Lahey held several technical and managerial positions with the General Electric Company, including overall responsibility for all domestic and foreign R&D programs associated with boiling water nuclear reactor thermal-hydraulic and safety technology. He has chaired several committees for the American Society of Mechanical Engineering, American Nuclear Society, American Institute for Chemical Engineering, American Society for Engineering Education, and NASA. His current research is funded by the Department of Energy's Naval Reactors Program, the Office of Naval Research, the National Science Foundation, the New York State Energy Research and Development Authority, Oak Ridge National Laboratory, and the Defense Advanced Research Projects Agency. He currently consults on nuclear reactor safety problems and the chemical processing of non-nuclear materials and is a member of the Board of Managers of PJM Interconnection, LLC. In 1994, he was elected to the National Academy of Engineering for his contributions to the fields of multiphase flow and heat transfer and nuclear reactor safety technology. In 1995, he became a member of the Russian Academy of Sciences-Baskortostan and he is a fellow of the American Nuclear Society and of the American Society of Mechanical Engineers. He has authored or co-authored over 300 technical publications, including 10 books or handbooks and 160 journal articles. Dr. Lahey holds a Ph.D. in mechanical engineering from Stanford University.

KATHLEEN R. MEYER has expertise in health physics and radiologic risk assessment. She is a principal of Keystone Scientific, Inc., and is currently involved in risk assessments for public health and the environment from radionuclides and chemicals at several U.S. Department of Energy sites. Other work includes an assessment of the interim radionuclide soil action levels adopted by the U.S. Department of Energy (DOE), the U.S. Environmental Protection Agency, and the Colorado Department of Health and Environment for cleanup at the Rocky Flats Environmental Technology Site. She has been a member of the National Council on Radiation Protection and Measurements Historical Dose Evaluation Committee. Dr. Meyer has authored or co-authored several peer-reviewed articles, including papers on cancer research, historical evaluation of past radionuclide and chemical releases, and risk assessment of radionuclides and chemicals. She holds a Ph.D. in radiological health sciences from Colorado State University.

FREDRICK J. MOODY is an expert thermal hydraulics and two-phase flow in nuclear power reactors. In 1999, he retired after 41 years of service at General Electric Company and 28 years as an adjunct professor of mechanical engineering at San Jose State University. Dr. Moody was the recipient of several prestigious career awards, including the General Electric Power Sector Award for Contributions to the State-of-the-Art for Two-Phase Flow and Reactor Accident Analysis. He has served as a consultant to the Nuclear Regulatory Commission's Advisory Committee on Reactor Safeguards, teaches thermal hydraulics for General Electric's Nuclear Energy Division, and continues to review thermal analyses for General Electric. Dr. Moody is a fellow of the American Society of Mechanical Engineers, which awarded him the George Westinghouse Gold Medal in 1980, and the Pressure Vessels and Piping Medal in 1999. He has also received prestigious career awards from General Electric and was elected to the Silicon Valley Engineering Hall of Fame. Dr. Moody was elected to the National Academy of Engineering in 2001 for pioneering and vital contributions to the safety design of boiling water reactors and for his role as educator. He has published three books and more than 50 papers. Dr. Moody holds a Ph.D in mechanical engineering from Stanford University.

TIMOTHY R. NEAL is an expert in weapons technology and explosives. He began his career at Los Alamos National Laboratory in 1967 and has led programs addressing weapon hydrodynamics, explosions inside structures and above ground, image analysis, and dynamic testing. He also has held several management positions within the Laboratory's nuclear weapons arena, including leadership of the Explosives Technology and Applications Division and of the Advanced Design and Production Technologies Initiative. He spearheaded Los Alamos' Stockpile Stewardship and Management Programmatic

Environmental Impact Statement and helped establish the U.S. Department of Energy's new Stockpile Stewardship Program. More recently, he has served as a senior technical advisor to the U.S. Department of Energy on nuclear explosive safety, and he has worked closely with the Pantex Plant for nuclear weapons production in Amarillo, Texas, in establishing a new formal basis for operational safety. Dr. Neal has received four DOE excellence awards, including one for hydrodynamics, and authored various technical papers and reports as well as one book on explosive phenomena. He holds a Ph.D. in physics from Carnegie-Mellon University.

LORING A. WYLLIE, JR. is an expert in structural engineering and senior principal of Degenkolb Engineers. His work has included seismic evaluations, analysis, and design of strengthening measures to improve seismic performance. He has performed seismic assessments and proposed strengthening solutions for several buildings within the U.S. Department of Energy weapons complex and for civilian buildings, some of which have historical significance. Mr. Wyllie's expertise is also recognized in several countries, including the former Soviet Union where he worked on an Exxon facility. Mr. Wyllie is a past president of the Earthquake Engineering Research Institute. His contributions to the profession of structural engineering were recognized by his election to the National Academy of Engineering in 1990 and his honorary membership in the Structural Engineers Association of Northern California. In recognition of Mr. Wyllie's expertise in concrete design and performance, the American Concrete Institute named him an honorary member in 2000. Mr. Wyllie also was elected an honorary member of the American Society of Civil Engineers in 2001. He holds a M.S. degree from the University of California, Berkeley.

PETER D. ZIMMERMAN is an expert in nuclear physics and terrorism. He is currently the chair of science and security and director of the Centre for Science & Security Studies at King's College in London. He previously served as the chief scientist of the Senate Foreign Relations Committee, where his responsibilities included nuclear testing, nuclear arms control, cooperative threat reduction, and bioterrorism. Previously, he served as science advisor for arms control in the U.S. State Department, where he provided advice directly to Assistant Secretary for Arms Control and the Undersecretary for Arms Control and International Security. His responsibilities included technical aspects of the Comprehensive Test Ban Treaty, biological arms control, missile defense, and strategic arms control. Dr. Zimmerman spent many years in academia as professor of physics at Louisiana State University. He is the author of more than 100 articles on basic physics as well as arms control and national security. His most recent publication is the monograph "Dirty Bombs: The Threat Revisited," which was published by the National Defense University in the Defense Horizons series. Dr. Zimmerman holds a Ph.D. in experimental nuclear and elementary particle physics from Stanford University and a Fil. Lic. degree from the University of Lund, Sweden. He is a fellow of the American Physical Society and a member of its governing council. He is a recipient of the 2004 Joseph A. Burton/Forum award for physics in the public interest.

C

TOUR OF SELECTED SPENT FUEL STORAGE-RELATED INSTALLATIONS IN GERMANY

On April 25-28, 2004, six committee members visited spent fuel storage-related installations in Germany. The following is a summary of some of the pertinent information obtained from that trip.

Several organizations and individuals worked with committee staff to make this trip possible. The committee would especially like to acknowledge Alfons Lührmann and William McConaghy of GNB/GNSI (Gesellschaft für Nuklear-Behälter, mbH/General Nuclear Systems, Inc.), who organized site visits; Klaus Janberg (STP engineering); Michael Sailer, chairman of RSK (Reaktorsicherheitskommission—reactor safety commission); Holger Broeskamp manager of GNS (Gesellchaft für Nuklear-Service, mbH—Germany's nuclear industry consortium) and his staff; Wolfgang Sowa, managing director of GNB (Gesellschaft für Nuklear-Behälter, mbH) and his staff; Florentin Lange of GRS (Gesellschaft für Anlagen- und Reaktorsicherheit, mbH); and Hubertus Flügge, vice-president of the RWE Power AG plants in Lingen and his staff, who allowed the committee to visit the reactor building and the site's spent fuel storage facility.

C.1 GERMAN COMMERCIAL NUCLEAR POWER PLANTS

Germany currently has 18 operating commercial nuclear power reactors at 12 sites. Approximately one-third of the reactors are boiling water reactors (BWRs) and two-thirds are pressurized water reactors (PWRs).

The design for PWR plants is illustrated schematically in FIGURE C.1. It consists of a dome-shaped reactor building constructed of reinforced concrete and a spherical inner containment structure constructed of steel. The reactor core, spent fuel pool, and steam generators are located within the inner containment. The emergency core-cooling systems are located outside the inner containment but within the reactor building.

The German BWR reactor building design is generally similar to a PWR. However, the spent fuel pool is outside the inner containment structure but within the reactor building. The reactor building is also a different shape (rectangular or cylindrical).

There are three generations of commercial nuclear power plants in Germany, each having increasingly thick walls:

- First-generation plants have reactor building walls that are less than 1 meter thick. There are four plants of this type.
- Second-generation plants have reactor building walls that are slightly more than 1 meter thick. There are five plants of this type.
- Third-generation plants have reactor building walls that are about 2 meters thick. There are nine plants of this type.[1]

[1] The committee subgroup visited one of these plants (the Lingen power plant) during its tour.

Some first- and second-generation plants have independent emergency systems in a bunkered building that contains some safety trains and a control room. These systems are capable of delivering water to the reactor after an accident or attack if the pipe systems within the reactor building survive.

Second- and third-generation plants were designed to withstand the crash of military fighter jets. Second-generation plants were designed to withstand the crash of a Starfighter jet at the typical landing speed. Third-generation plants were designed to withstand the crash of a Phantom jet at the typical cruising speed. This is considered to be part of the "design basis threat" for nuclear power plants in Germany. This information on the design basis threat has been made available to the public by the German government.

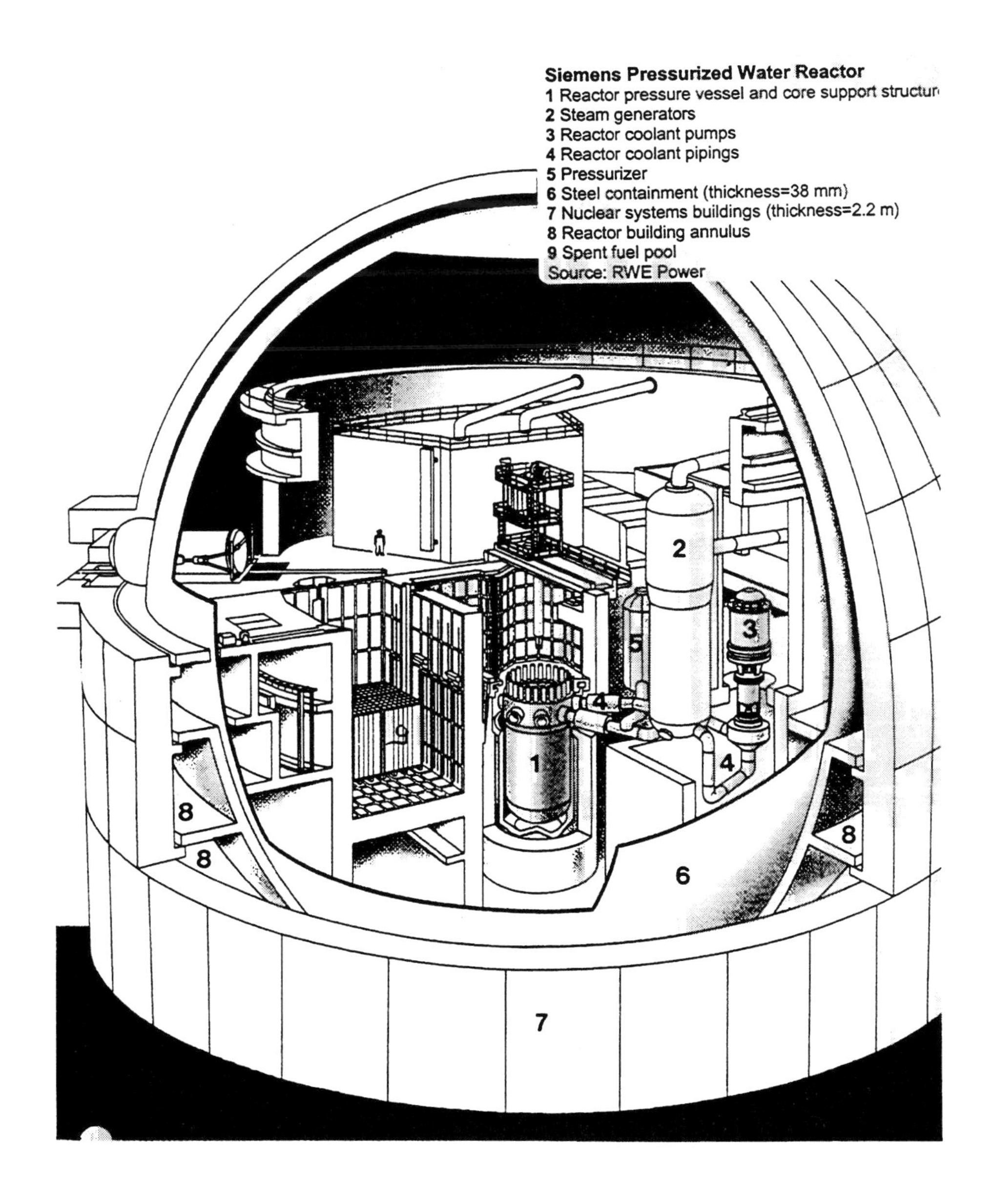

FIGURE C.1 Schematic illustration of the Lingen PWR power plant, a third-generation power plant design. SOURCE: RWE Power.

Plant operators must show that of the four safety trains (each train contains 50 percent of the safety system) at the plant, at least two will survive such a crash. The crash parameters (e.g., aircraft type, speed, and angle) have been established by RSK. The crash parameters have been published and the public knows about them. Each plant must perform an independent analysis of each reactor building. Sometimes two separate analyses have to be provided for the same site if there are two or more reactors with different designs.

In 1998, the German government decided to phase out nuclear energy. Commercial nuclear plants will be allowed to generate an agreed-to amount of electricity before shutdown. Currently, the Lingen and the Neckarwestheim-2 plants have the highest remaining electricity production allowance and will be shut down in 2021 or 2022, should no revision of this political decision be implemented.

C.2 SPENT FUEL STORAGE

Until recently, all spent fuel at German plants was stored in the reactor pools until it could be sent to Sellafield (U.K.) or La Hague (France) for reprocessing. In the 1980s, plants began to re-rack their spent fuel pools to increase storage capacities (the older German nuclear plants were designed to contain one full reactor core plus one third of a core). Regulators became concerned that the emergency cooling systems were not sufficient to handle the increased heat loads in spent fuel pools from this re-racking. Some plants added additional cooling circuits to address this concern. Only one power plant (an older plant at Obrigheim) has wet interim pool storage in a bunkered building.

A discussion of alternative spent fuel storage options began in 1979. A reprocessing plant had been proposed at Gorleben that would have had several thousand metric tons of pool storage. The German government concluded that while there were no major technical issues for reprocessing, wet fuel storage was a potential problem because cooling systems could be disrupted in a war. GNS decided to shift from wet to dry storage for centralized storage facilities.

There are two centralized storage facilities in Germany: Gorleben and Ahaus. Gorleben is designed to store vitrified high-level waste from spent fuel reprocessing and spent fuel from commercial power reactors. Ahaus is designed to store spent fuel from test reactors and other special types of fuel. Ahaus currently stores 305 casks of reactor fuel from the decommissioned Thorium High Temperature Reactor, three casks of PWR spent fuel from the Neckarwestheim site, and three casks of BWR spent fuel from the Gundremmingen site. The latter shipment produced large public demonstrations and required the deployment of 35,000 police officers to maintain security.

At the end of 2001, the German utility companies and the German federal government agreed to avoid all transport of spent fuel in Germany because of intense public opposition. The German government recently passed a law making it illegal to transport spent nuclear fuel to reprocessing plants in France and the United Kingdom after June 30, 2005. However, there is no legal restriction concerning the transport of spent fuel from power reactors to other destinations (e.g., to dry storage facilities). The government and power plant operators have negotiated an agreement to develop dry cask storage facilities at each of the 12 nuclear power plant sites to avoid the need for offsite spent fuel transport.

These dry cask storage facilities are to be constructed by 2006. They are licensed to store fuel for 40 years.

There are three dry cask storage facility designs in Germany:

1. WTI design: The walls and roof are constructed of 80 and 50 centimeters, respectively, of reinforced concrete.
2. STEAG design: The walls and roof are constructed of 1.2 and 1.3 meters, respectively, of reinforced concrete. This design is used at the Lingen Nuclear Power Plant dry storage facility visited by the committee (FIGURE C.2).
3. GNK design: This is a tunnel design and is under construction at the Neckarwestheim nuclear power plant.

The use of reinforced concrete in these facilities was originally intended for radiation protection and structural support, not for terrorist attacks.

In 1999, RSK issued guidelines for dry storage, which were released in 2001 (RSK, 2001). Licensing a dry storage facility in Germany requires several safety demonstrations and analyses. As part of the licensing procedures for a storage facility, the license applicant must do independent calculations that demonstrate how the building features meet the safety standards and the design basis threat. This threat includes an armed group of intruders and the impact of a Phantom 2 military jet. It also includes a shaped charge. The scenario of a deliberate crash of a large civilian airplane has been considered and analyzed as part of the recent licensing of onsite dry storage facilities but is not established as part of the design basis threat. There are public hearings during which the license applicant explains the safety features of the storage facility. The public is aware of the design basis threat, and it is provided with the results of the analysis but not with the details.

FIGURE C.2 Dry cask spent fuel storage building at the Lingen Nuclear Power Plant.
SOURCE: RWE Power.

There are six temporary (i.e., five- to seven-year) storage facilities in use at reactor sites until these dry cask storage facilities become available. The casks in these temporary storage facilities are stored horizontally and are protected by concrete "garages" designed to withstand the impact of a Phantom military jet.

Spent commercial fuel is stored in CASTOR® casks (FIGURE C.3) that were originally designed and developed by the German utility-owned company GNB.[2] These casks can store either PWR or BWR spent fuel assemblies. The design consists of a ductile cast iron cylindrical cask body with integral circumferential fins machined into the outer surface to maximize heat transfer; inside, the spent fuel assemblies are inserted in a borated stainless steel basket. The cask has a double-lid system that is protected by a third steel plate. The cask complies with the international regulations of the International Atomic Energy Agency (IAEA) as a type B(U) package.

Spent fuel is typically cooled for five years in a pool before it is put in dry cask storage; some other custom-made cask designs can hold fuel that has been cooled for shorter (minimum two years) or longer times depending on the fuel characteristics and fuel burn-up. Current fuel burn-ups in Germany (52 to 55 gigawatt-days per metric ton) are similar to those in the United States.

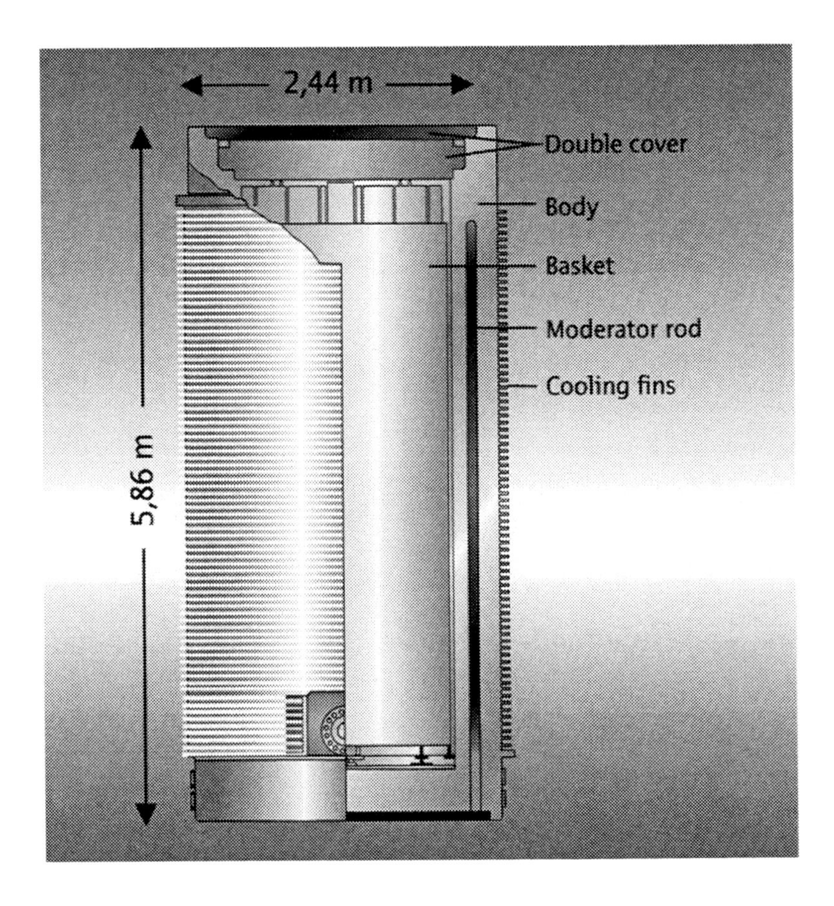

FIGURE C.3 Typical features of a CASTOR cask used at the Lingen Nuclear Power Plant.
SOURCE: RWE Power AG Lingen Nuclear Power Plant.

---

[2] Gesellschaft für Nuklear-Behälter, mbH.

C.3 RESPONSE TO THE SEPTEMBER 11, 2001, TERRORIST ATTACKS IN THE UNITED STATES

The September 11, 2001, terrorist attacks on the United States caused the German government to reassess the security of its nuclear power plants and spent fuel storage facilities. RSK held meetings starting in October 2001 to discuss the implications of the September 11 attacks for German commercial nuclear power plants. It issued a short statement recommending that an analysis be carried out on each plant to assess its vulnerability to September 11-type attacks. These analyses have not yet been undertaken. Plant operators assert that terrorist attacks are a general risk of society and should be treated like attacks on other infrastructure (e.g., chemical facilities). The Länder (state) governments, which are responsible for licensing commercial power plants in Germany, do not require these analyses. RSK recommended that the federal government develop a checklist for such an analysis, but this also has not been done.

A general analysis of the impact of the different civilian aircraft on commercial nuclear plants was requested by BMU[3] and has been carried out by GRS.[4] The result of the discussions between RSK and BMU on the basis of this report was that plant specific sensitivity analyses are needed. GRS was also involved in the framing of the recent German licensing process in the analysis of the consequences of civilian aircraft attacks on STEAG- and WTI-design spent fuel storage facilities using three sizes of aircraft (ranging from Airbus A320- to Boeing 747-size aircraft).

C.4 TESTS ON GERMAN CASKS

The casks that are used in German dry cask storage facilities have been subjected to several tests that simulate accidents and terrorist attacks. The following types of tests were performed on these casks or cask materials.

Airplane crash test simulations with military aircraft (Phantom type) are part of the licensing requirements for both casks and storage facilities. Between 1970 and 1980 a number of tests on storage casks were carried out at the Meppen military facility in Germany. A one-third scale model of a GNB cask was used to simulate the impact of a turbine shaft of a military aircraft using a hollow-tube projectile. Two different impact orientations were used: perpendicular to upright cask body (lateral impact) and perpendicular to center of lid system. The projectile completely disintegrated in the test, but the cask sustained only minor damage.

The jet aircraft tests were carried out because of safety concerns, but after September 11, 2001, intentional crashes of aircraft also were considered. Investigations by BAM (Bundesanstalt für Materialforschung und -prüfung) and GRS concluded that CASTOR-type casks would maintain their integrity when intentionally hit by a commercial aircraft.

---

[3] Bundesministerium für Umwelt, Naturschutz and Reaktorsicherheit (Federal Ministry for Environment, Nature Protection, and Nuclear Safety and Security).

[4] Gesellschaft für Anlagen- und Reaktorsicherheit (GRS), mbH (Company for Installation and Reactor Safety). GRS is Germany's main research institution on nuclear safety. It is an independent, nonprofit organization, founded in 1977, and has about 450 employees. GRS funds its work through research contracts. Some have compared GRS to Sandia National Laboratories in the United States.

Other types of terrorist attacks have been a long-standing concern to the German government because of terrorism activities in Europe in the 1970s and 1980s. A series of tests simulating terrorist attacks on casks were done in Germany, France, the United States (for the German government), and Switzerland (for the Swiss government). Additional tests may have been done that are not publicly acknowledged.

In 1979-1980 at the German Army facility in Meppen, a "hollow charge" (i.e., shaped charge) weapon was fired at a ductile cast iron plate and fuel assembly dummy to simulate a CASTOR cask. The cask plate was perforated but release fractions from the fuel assembly were not examined. From this experiment, the German government concluded that the wall thickness of the cask should not be less than 300 millimeters.

Other tests were carried out at the Centre d'Etude de Gramat in France in 1992 on behalf of the Germany Federal Ministry of Environment, Nature Protection and Nuclear Safety (BMU) (Lange et al., 1994). These tests involved shaped charges directed at a CASTOR cask (type CASTOR IIa, the cask was one third of the regular length) filled with nine fuel element dummies with depleted uranium. The fuel rods were pressurized to 40 bars to simulate fuel burn-up, but the cask interior was at atmospheric pressure or at reduced pressure of 0.8 bar. The shaped charge perforated the cask and penetrated fuel elements. This damaged the fuel and resulted in the release of fuel particles from the cask.

These particles were collected, and their particle size distribution was measured. About 1 gram of uranium was released in particles of less than 12.5-microns aerodynamic diameter, and 2.6 grams of uranium were released in particles with a size range between 12.5 and 100 microns. If the pressure inside the cask was reduced to 0.8 bar (to simulate the conditions during interim storage of spent fuel in Germany), the releases were reduced by two-thirds: 0.4 gram for particle sizes less than 12.5 microns and about 0.3 gram for particles between 12.5 and 100 microns.

In 1998, a demonstration was carried out at the Aberdeen Proving Ground in the United States using an anti-tank weapon on a CASTOR cask. The purpose of this demonstration was to show that a concrete jacket on the exterior of the cask could prevent perforation. The weapon was first fired at the cask without the jacket. It perforated the front wall of the cask. The concrete jacket was effective in preventing perforation of the cask. Committee members saw a specimen of this cask at the GNB workshop (see FIGURE C.4).

Also in 1999, explosion of a liquid gas tank next to a cask was performed by the German BAM (Federal Office of Material Research and Testing) to study the effect of accidents involving fire or explosions in the vicinity of the cask during transportation or storage. The gas tank and the CASTOR cask were initially about 8 feet (2.5 meters) apart. Explosion of the tank generated a fire ball 330 to 500 feet (100 to 150 meters) in diameter. The explosion projected the cask 23 feet (7 meters) away and tilted it by 180 degrees, causing it to hit the ground on the lid side. Examination after the explosion showed no change in the containment properties of the lid system.

FIGURE C.4 Section of a CASTOR cask showing the perforation made by a shaped charge at the Aberdeen Proving Ground. SOURCE: Courtesy of GNB/GNSI.

REFERENCE

Lange, F., G. Pretzsch, J. Döhler, E. Hörmann, H. Busch, and W. Koch. 1994. Experimental Determination of UO_2-Release from a Spent Fuel Transport Cask after Shaped Charge Attack. 35th INMM Annual Meeting Proceedings (Naples, Florida). Vol. 23, pp. 408-413.

RSK (Reaktorsicherheitskommission). 2001. Safety-Related Guidelines for the Dry Interim Storage of Spent Fuel Elements in Storage Casks. Recommendation of the Commission on Reactor Safety. April 5. Available at *http://www.rskonline.de/Download/Leitlinien/English/RSK-GUIDELINES-DRY-INTERIM-STORAGE.pdf.*

D

HISTORICAL DEVELOPMENT OF CURRENT COMMERCIAL POWER REACTOR FUEL OPERATIONS

There are 103 commercial power reactors operating in the United States at this time. Almost all of them are operating with spent fuel pools that are too small to accommodate cumulative spent fuel discharges. This short appendix was prepared to provide a historical background for power reactor fuel operations and pool and dry-cask storage of spent fuel.

D.1 DESIGN FOR A CLOSED FUEL CYCLE

The first large generation of commercial reactors in the United States were almost all light water reactors (LWRs), that is, nuclear reactors that use ordinary water to cool the core and to moderate the neutrons emitted by fission. The hydrogen atoms in the water coolant moderate, or slow down the fission-emitted neutrons to an energy level that is more likely to cause fission when the neutron strikes a fissile atom. These reactors were designed, developed, and licensed in the 1960s and 1970s, although many were not completed until the 1980s. Their design power output increased rapidly, as it did for non-nuclear power plants, in order to achieve economies of scale. Thus, the earlier plants in this generation were designed to produce 500-900 megawatts of electrical power (MWe) while later units increased to 1000-1200 MWe. The number of LWRs built and ordered by the U.S. industry began to approach 200. All of these plants were being designed for a closed fuel cycle, that is, for the uranium oxide fuel, enriched to 2-5 percent uranium-235, to be loaded and "burned" to a level of 20-30 gigawatt-days per metric ton of uranium (GWd/MTU), then reprocessed in commercial plants to separate the still usable fissionable, or fissile, materials in the spent fuel from the radioactive waste. The reprocessing plants would recover the fissile plutonium-239 formed from uranium-238 during reactor operations and residual fissile uranium-235 for use as fuel in LWRs and later in breeder reactors (USNRC, 1976).

By the mid-1970s commercial reprocessing plants were built, under construction, or planned in New York, Illinois, South Carolina, and Tennessee, with a combined projected capacity to reprocess more than 6000 MTU of spent fuel per year. For comparison, a large LWR discharges about 20 MTU of spent fuel at a refueling. By this time the price of fresh uranium was dropping and the cost of fuel reprocessing made it difficult for recycle fuel to compete with fresh fuel. Also, there was controversy about the risk of fissile material diversion if recycled plutonium was moved in commercial traffic. Both existing fuel reprocessing plants withdrew from licensing for technical reasons and then, on April 7, 1977, President Carter issued a policy statement that "we will defer indefinitely the commercial reprocessing and recycling of the plutonium produced in the U.S. nuclear power programs." The statement went on to say: "The plant at Barnwell, South Carolina, will receive neither federal encouragement nor funding for its completion as a reprocessing facility." After consultation with the White House, the U.S. Nuclear Regulatory Commission (USNRC) terminated its Final Generic Environmental Statement on the Use of Recycled Plutonium in Mixed Oxide Fuel in Light-Water Cooled Reactors (GESMO) proceedings.

Thus, the U.S. nuclear industry was immediately changed from a closed fuel cycle, with recycle, to an open or once-through fuel cycle with the fuel loaded into the reactor in several consecutive locations to obtain maximum economic use of the fuel before it was finally removed as waste. The USNRC changed the legal definition of high-level radioactive waste to include the high-level waste from both nuclear fuel reprocessing and spent nuclear fuel.

For this study, the significance of this closed fuel cycle design is that this entire generation of more than 100 reactors was designed with small spent fuel pools, relying on prompt shipment away from the reactor to the reprocessing plant to make room for later discharges of spent fuel. Early spent fuel shipping casks were being designed with active cooling systems to support shipment of fuel less than a year out of the reactor to a reprocessing plant. BOX D.1 discusses the spent nuclear fuel at reprocessing plants. Supplementary wet and dry storage systems had to be developed to receive the older spent fuel to make room for fresh spent fuel from the reactor. Many plants had to remove and modify the storage racks in their spent fuel pools to accommodate more spent fuel in the pool itself until licensed supplementary systems were available.

D.2 RETRENCHMENT OF U.S. REACTOR PLANS

As noted in Section D.1, in the 1970s the United States was building reactors at a high rate. Then, in the late 1970s, three factors produced a retrenchment in power reactor plans: rising interest rates, reversal of the U.S. fuel reprocessing policy, and the Three Mile Island-2 accident.

D.2.1 Effect of Interest Rates

Commercial power reactors have characteristically high initial capital costs. The regulated public utilities have had to raise the capital with various debt instruments; to build, license, and operate the finished plant for a time before it can be declared commercial; and to change the electricity rates charged consumers to retire the debt on the capital cost. The soaring interest rates in the United States during the late 1970s drove the costs of new nuclear plants that were under construction to extreme heights. This, combined with slackening demand for electricity, led to the cancellation of many plants, some even in advanced stages of construction.

D.2.2 Effect of Reversal of U.S. Fuel Reprocessing Policy

President Carter enunciated a change in U.S. policy for reprocessing of spent nuclear fuel in early 1977. Those reactors then operating and those under construction had to begin modifying their reactor fuel cycle design to go from the closed (reprocessing) cycle to a “once-through” fuel cycle. This induced the designers to go to higher levels of uranium-235 enrichment in the new fuel, but still within the 5 percent licensing limit. It also induced the designers to revise the core loading and operating plans in order to burn or use the fissile content of the fuel to the greatest extent economically possible since the fissile residue could not be retrieved by reprocessing. As a result, spent fuel burnup levels rose to levels that are now almost double the 20-30 GWd/MTU characteristic of the original closed fuel cycle. This results in an increase in the decay-heat power of the spent fuel assembly by the time it is put into the spent fuel pool.

BOX D.1 Spent Fuel at Nuclear Fuel Reprocessing Plants

Up until the mid-1970s the commercial nuclear industry was expected to operate several nuclear fuel reprocessing plants to recover fissile plutonium from virtually all of the commercial spent fuel from U.S. reactors. These plants would use aqueous reprocessing methods developed by the Atomic Energy Commission (AEC). The recovered plutonium was to be used as mixed oxide fuel (PuO_2 and UO_2) in water reactors and, later, as fuel in breeder reactors. Each reprocessing plant had one or two storage pools to receive and store the fuel temporarily until it was reprocessed. No long-term storage of the spent fuel from commercial reactors was planned. Only two commercial reprocessing sites have received spent fuel, West Valley, New York, and G.E.-Morris, Illinois.

The first commercial reprocessing plant began operations by the Nuclear Fuel Services Company on a site in West Valley, New York, owned by the State of New York. The State of New York licensed a low-level radioactive waste disposal site adjacent to the reprocessing plant. The West Valley plant had a reprocessing capacity of about 1 metric ton of uranium (MTU) per day. It operated at reduced capacity because there was not yet much commercial spent fuel to reprocess. In fact, about half of the spent fuel reprocessed there was from the last in the series of plutonium production reactors, the N-Reactor, at the AEC site in Hanford, Washington. This spent fuel was provided to the West Valley plant to keep it working in the early days when little commercial spent fuel was available. The West Valley plant suspended operations in 1972 in order to expand its capacity to about 3 MTU per day. The work and the re-licensing effort went on until 1976 when the company withdrew its application for the new license and terminated reprocessing operations. The U.S. Department of Energy (DOE) took over the task of high-level radioactive waste retrieval and decommissioning under the West Valley Demonstration Project Act of 1980. About 137 MTU of commercial spent fuel remaining in the cooling pool was returned to its owners (USNRC, 1987). In 2003 the last of this spent fuel, about 25 MTU in two shipping casks, was shipped to the DOE-Idaho National Lab where it remains in dry storage in those casks.

The General Electric Company built a nuclear fuel reprocessing plant at Morris, Illinois, near the Dresden Nuclear Power Station. The plant was expected to reprocess 3 MTU per day. When the G.E.-Morris plant was in its final testing in 1975, the company determined that its performance would not be acceptable without extensive modifications. The request for a reprocessing plant operating license was withdrawn and the plant was licensed only to possess the spent nuclear fuel that it was under contract to reprocess. After modifying the storage system in its below-grade pool to hold more spent fuel, G.E.-Morris has received and stores 700 MTU of spent fuel for various owners.

Power reactors are refueled, and spent fuel is discharged to the storage pool, every one to two years. The decay-heat power of recently discharged spent fuel dominates the heat load of all the spent fuel in the pool, both freshly discharged and old, since the decay heat from a spent fuel assembly decreases by one to two orders of magnitude in the first year after it is removed from the reactor. Increasing the capacity of the spent fuel pool by re-racking, that is, modifying the storage racks to provide for closer spacing of the fuel assemblies,[1] allows older fuel to be accumulated in the pool rather than being removed for

[1]The capacity of spent fuel pools has typically been increased by replacing the original storage racks with racks that hold the spent fuel assemblies closer together. The fuel assembly channels in these

shipment or dry storage. Re-racking can make it more difficult to cool the freshly discharged fuel if there is catastrophic loss of the fuel pool water.

D.2.3 Effect of the Three Mile Island Accident

The final factor driving the retrenchment of the nuclear power industry was the Three Mile Island-2 (TMI-2) accident that occurred on March 28, 1979, in Pennsylvania (Walker, 2004). In that accident a small failure in the reactor coolant system was compounded by operator errors to result in catastrophic damage; a partial core melt occurred. The inability of the operators to understand and control the events, and the confusion among the state, the USNRC, and other responsible agencies about public protection had a devastating effect on public trust in the safety of nuclear power. The USNRC escalated safety requirements after the TMI-2 accident. These new requirements substantially modified the operation of licensed plants, delayed completion of new plants, and further increased their construction costs. The accident also resulted in the retrenchment of nuclear power in the 1980s and led to the cancellation of many plants, decommissioning of some plants, and the sale of some plants to other owners. The fleet of operating U.S. reactors was reduced to the presently operating 103 described here.

D.3 COMMERCIAL POWER REACTORS CURRENTLY OPERATING IN THE UNITED STATES

All of the commercial power reactors operating in the United States are light water reactors. BOX D.2 describes the LWRs that are currently operating in the United States.

D.3.1 Pressurized Water Reactors

About two-thirds of the U.S. reactors are pressurized-water reactors (PWRs), dual-cycle plants in which the primary cooling water is kept under a pressure of about 2000 pounds per square inch absolute (psia) as it circulates to remove fission and decay heat from the reactor fuel in the core and carry that energy to the steam generators, to generate steam in the lower-pressure secondary loop. The reactor, primary loop piping, and steam generators are all located in the containment structure; the steam lines penetrate the containment carrying the steam to the turbine to generate electrical power.

About one-third of the U.S. reactors are boiling-water reactors (BWRs), single-cycle plants in which the primary coolant of the reactor core is operated at about 1000 psia as it recirculates within the reactor core. The fission and decay heat generated in the core cause a substantial amount of the reactor coolant water to boil into steam that passes out directly from the reactor pressure vessel to the turbine-generator system. Plant differences stem initially from the different designs of the nuclear steam system supplier, the different designs of the architect-engineers that built the plants, and the owners that often specified additional modifications.

---

replacement racks typically have solid metal walls with neutron-absorbing material for nuclear safety reasons. This configuration inhibits water or air circulation more than the earlier configuration.

BOX D.2 U.S. Nuclear Power Plants

In the United States, 32 utility companies are licensed to manage the 103 operating reactors. There are also 27 shutdown reactors in storage or decommissioning. These reactors are situated at 65 nuclear power plant sites across the United States; a plant site may have 1, 2, or 3 reactors.

The fleet of 103 operating reactors in the United States is composed of the following:

- 69 pressurized water reactors (PWRs) and
- 34 boiling water reactors (BWRs).

The containment design for PWRs is divided into dry (56 reactors), ice condenser (9 reactors), and sub-atmospheric (4 reactors) containments. Among the BWR containment designs, 22 reactors are of design type Mark I, 8 of Mark II, and 4 of Mark III.

The PWRS operating in the United States were designed by three different nuclear steam system suppliers; Westinghouse Electric, Combustion Engineering, and Babcock & Wilcox. Most PWRs have what are called large dry containments, that is, containment structures of about 2 million cubic feet volume that can absorb the rapid release of steam and hot water from a postulated rupture of the primary coolant system without exceeding an internal pressure of about 4 atmospheres. FIGURE D.1 illustrates a PWR in a large dry containment. Some PWR containments are essentially as large but use ventilation fans to maintain the initial containment pressure mildly sub-atmospheric to provide an additional pressure margin. Finally, one set of nine Westinghouse PWRs uses ice-condenser containment structures, in which the containment has about the same pressure capability but is smaller, relying on massive baskets of ice maintained in the containment to condense steam releases and mitigate the pressure surge.

D.3.2 Boiling Water Reactors

The BWRs in operation today were designed by the General Electric Company. They all use pressure suppression containments, two-chamber systems with the reactor located in a dry well that is connected to a wet well containing a large pool of water.

In the event of a rupture of the reactor system in the dry well, the steam and hot water released are channeled into the water in the wet well, condensing and cooling the steam to mitigate the pressure surge. BOX D.2 lists the three successive generations of BWR containment design, and the number of each still operating. FIGURE D.2 illustrates three types of BWR containments: Mark I, Mark II, and Mark III. The Mark I containment is the most common type with 22 in operation. The reactor pressure vessel, containing the reactor core is located in a dry well of the containment in the shape of an inverted incandescent light bulb.

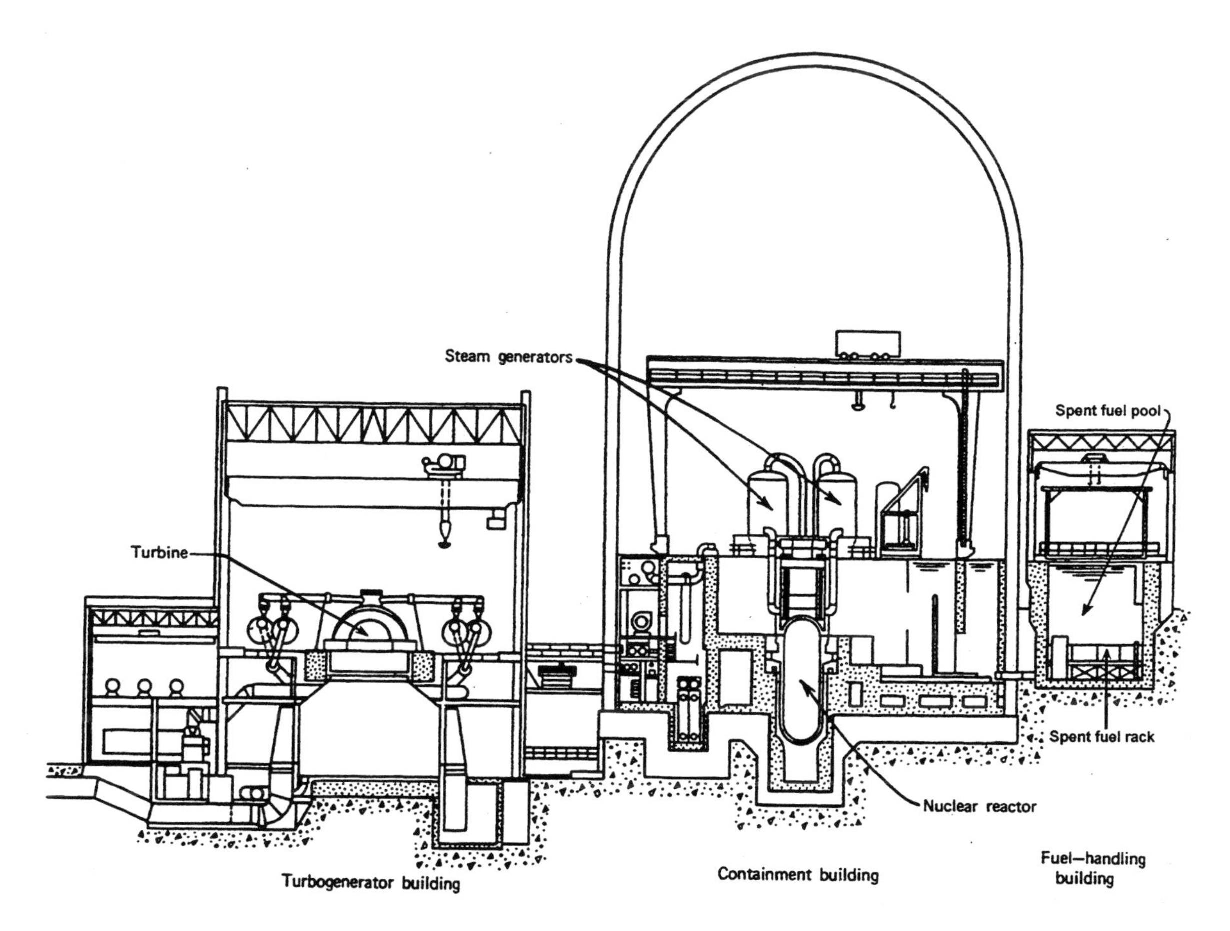

FIGURE D.1 A PWR in a large dry containment. SOURCE: Modified from Duderstadt and Hamilton (1976, Figure 3-4).

The dry well is connected by large ducts to the wet well, a large toroidal (i.e., doughnut-shaped) part of the containment that is partially filled with water. Gas and steam releases from an accident in the dry well would be passed through the connecting ducts into the water in the wet well, cooling the gas and condensing the steam to mitigate the accident pressure rise in the containment. The containment building Mark II BWR is similar to the Mark I except that in the Mark II containment the conical dry well is directly above the cylindrical wet well. Nine Mark II reactors are still operating in the United States. In the Mark III, the dry well around the reactor vessel is vented to the top of a cylindrical wet well that surrounds it.

Four Mark III BWRs are currently operating. The entire dry well-wet well system is contained within a large steel containment shell and a concrete shield building.

D.3.3 Reactor Fuel and Reactor Control

TABLE D.1 presents the range of dimensions and weights for a wide variety of the LWR fuel assemblies used in the operating reactors. The spent fuel pools and the dry storage systems used at a reactor must be tailored to the specific fuel design for that reactor.

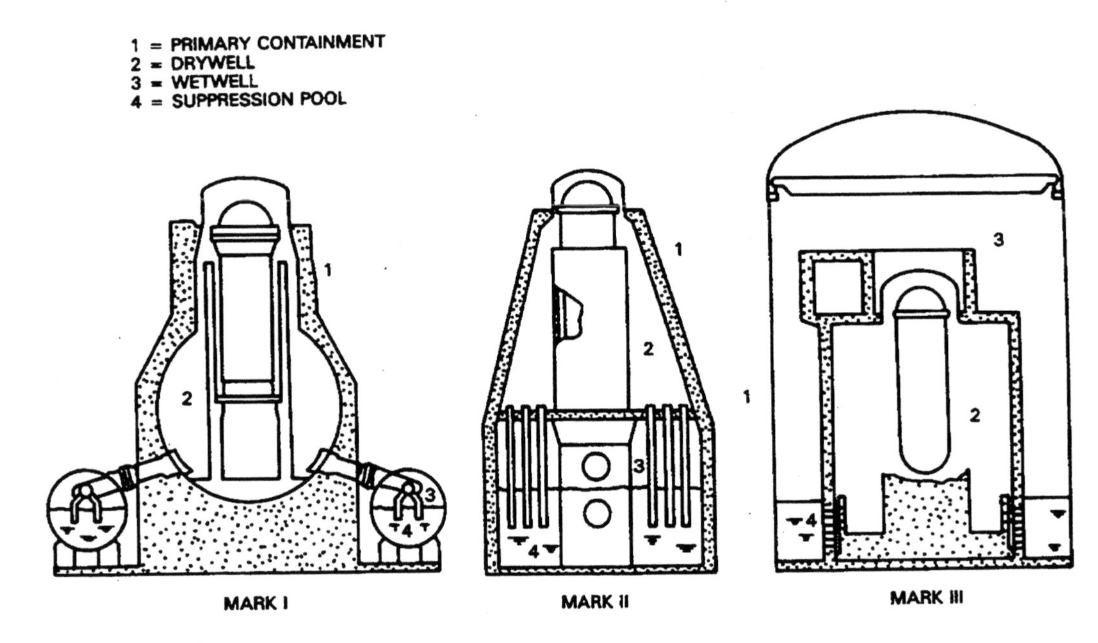

FIGURE D.2 Three types of BWR containment system: Mark I, Mark II, and Mark III. SOURCE: Modified from Lahey and Moody (1993, Figure 1-9).

The fission process is controlled by the reactor operators through the use of neutron-absorbing materials. The primary control is an array of control rods or blades that can be withdrawn from the core to the degree needed. In the PWRs, the control rods are moved within selected empty tubes within the assembly. In the BWRs, cruciform (cross-shaped) control blades are moved across the faces of the fuel assembly, typically narrower than those in a PWR fuel assembly. Reactor fuel designers also use burnable poisons within the fuel assembly to control the fission process. These poisons are placed in appropriate amounts within the fuel assembly so that they burn away, making the fuel assembly more reactive, as the continued fission process is making it less reactive. PWRs also use neutron control by dissolving neutron-absorbing sodium borate in the reactor coolant, gradually lowering the concentration from the peak after refueling to the minimum before the next refueling.

REFERENCES

American Nuclear Society. 1988. Design Criteria for an Iindependent Spent Fuel Storage Installation (Water Pool Type): An American National Standard. ANSI/ANS-57.7-1988. American Nuclear Society. LaGrange Park, Illinois.

Duderstadt, J. J. and L. J. Hamilton. 1976. Nuclear Reactor Analysis. John Wiley & Sons. New York.

Lahey, R. T. and F. J. Moody. 1997. The Thermal Hydraulics of a Boiling Water Nuclear Reactor. Second Edition. American Nuclear Society. La Grange Park, Illinois.

USNRC (U.S. Nuclear Regulatory Commission). 1976. Final Generic Environmental Statement on the Use of Recycled Plutonium in Mixed Oxide Fuel in Light-Water Cooled Reactors (GESMO). NUREG-0002. Washington, DC.

USNRC. 1987. Case Histories of West Valley Spent Fuel Shipments. NUREG/CR-4847. January. Washington, D.C.

Walker, J. S. 2004. Three Mile Island: A Nuclear Crisis in Historical Perspective. University of California Press. Berkeley, California.

TABLE D.1 Range of Dimensions and Weights for Light Water Reactor Fuel Assemblies Used in Operating Reactors in the United States.

Physical Characteristics of Typical LWR Fuel Assemblies

Reactor Type	BWR	BWR	PWR	PWR	PWR	PWR	PWR	PWR	PWR	PWR	PWR	PWR
Fuel Designer	GE	GE	B&W	B&W	GE	GE	W	W	W	W	W	W
Fuel Rod Array	7x7	8x8	15x15	17x17	14x14	16x16	14x14	14x14	15x15	15x15	17x17	17x17
Active Fuel Length (in.)	144	144	144	143	137	150	120	144	121	144	144	168
Nominal Envelope (in.)[2]	5.438	5.47	8.536	8.536	8.25	8.25	7.763	7.763	8.449	8.426	8.426	8.426
Fuel Assembly Length (in.)	176	176	166	166	157	177	137	161	137	160	160	—
Weight (lbs.)	600	600	1,516	1,502	581 kg	—	501 kg	573 kg	594 kg	654 kg	665 kg	—
Fuel Rod												
Number	49	63	208	264	164	224-236	180	179	204	204	264	264
Length (in.)	163	—	153	—	147	161	127	152	127	152	152	—
Pitch, Square (in.)	0.738	0.640	0.568	0.501	0.580	0.506	0.556	0.556	0.563	0.563	0.496	0.496
O.D. (in.)	0.570	0.493	0.430	0.379	0.440	0.382	0.422	0.422	0.422	0.422	0.374	0.360
Clad Thickness (mils.)	35.5	34	26.5	23.5	26	25	16.5	24.3	16.5	24.3	22.5	22.5
Clad Material	Zr 2	Zr 2	Zr 4	Zr 4	Zr 4	Zr 4	sst	Zr 4	sst	Zr 4	Zr 4	Zr 4
Pellet O.D. (in.)	0.488	0.416	0.370	0.3232	0.3795	0.325	0.3835	0.3659	0.3835	0.3659	0.3225	0.3088
Pellet Length (in.)	—	—	—	0.375	0.650	0.390	0.600	0.600	0.600	0.600	0.530	0.530
Gap, Radial (mils.)	5.5	4.5	3.5	3.1	4.3	3.5	2.8	3.8	2.8	3.8	3.3	3.3
Density (STD)	—	—	92.5-95.0	93.5-95.0	93.0-95.0	94.75	93.0-94.0	92.0	93.0-94.0	92.0	95.0	95.0
Poison	Gd_2O_3	Gd_2O_3	None	None	B_4C/Al_2O_3	B_4C/Al_2O_3	—	—	—	—	—	—
Nonfueled Rods												
Number	0	1	17	25	6	6	16	17	21	21	25	25
Material	—	Zr 2	Zr 4	Zr 4	Zr 4	Zr 4	304 sst	Zr 4	304 sst	Zr 4	Zr 4	Zr 4
Spacer Grids												
Number	7	7	8	8	8	12	—	—	—	—	—	—
Material	Inconel X	Inconel X	Inconel 718	Inconel 718	Zr 4	Zr 4	—	—	—	—	—	—

SOURCE: American Nuclear Society (1988).

E

GLOSSARY

Actinide: Any of a series of chemically similar radioactive elements with atomic numbers ranging from 89 (actinium) through 103 (lawrencium). This group includes uranium and plutonium.

Alpha particle: Two neutrons and two protons bound as a single particle (a helium nucleus) emitted from certain radioactive isotopes when they undergo radioactive decay.

Bare-fuel cask: See *Cask*.

Beta particle: A charged particle consisting of a positron or electron emitted from certain radioactive isotopes when they undergo radioactive decay.

Beyond-design-basis accidents: Technical expression describing accident sequences outside of those used as design criteria for a facility. Beyond-design-basis accidents are generally more severe but are judged to be too unlikely to be a basis for design.

Boiling water reactor (BWR): A type of nuclear reactor in which the reactor's water coolant is allowed to boil to produce steam. The steam is used to drive a turbine and electrical generator to produce electricity.

Burn-up: Measure of the number of fission reactions that have occurred in a given mass of nuclear fuel, expressed as thermal energy released multiplied by the period of operation and divided by the mass of the fuel. Typical units are megawatt-days per metric ton of uranium (MWd/MTU) or gigawatt-days per metric ton of uranium (GWd/MTU).

Canister-based cask: See *Cask*.

Cask: Large, typically cylindrical containers constructed of steel and/or reinforced concrete that are used to store and/or transport spent nuclear fuel. Casks designed for storage of spent nuclear fuel can be of two types: "bare-fuel" or "canister-based." In bare-fuel casks, spent fuel is stored in a fuel basket surrounded by a heavily shielded and leak-tight container. In canister-based casks, the fuel is enclosed in a leak-tight steel cylinder, called a canister, which has a welded lid. The canister is placed in a heavily shielded cask overpack. Casks can be single-, dual-, or multiple-purpose, indicating that they can be used, respectively, for storage (also called storage-only casks), for storage and transportation, and for storage, transportation, and geologic disposal. There are no true multi-purpose casks for spent fuel currently available on the market.

Cesium-137: Radioactive isotope that is one of the products of nuclear fission.

Chain reaction: A series of fission reactions wherein the neutrons released in one fission event stimulate the next fission event or events.

Cladding: Thin-walled metal tube that forms the outer jacket of a nuclear fuel rod. It prevents corrosion of the nuclear fuel and the release of fission products into the coolant. Zirconium alloys (also called *zircaloy*, see below) are common cladding materials in commercial nuclear fuel.

Conduction: In the context of heat transfer, the transfer of heat within a medium through a diffusive process (i.e., molecular or atomic collisions).

Containment structure: A robust, airtight shell or other enclosure around a nuclear reactor core to prevent the release of radioactive material to the environment in the event of an accident.

Convection: Heat transfer by the physical movement of material within a fluid medium.

Cooling time: The amount of time elapsed since spent fuel was discharged from a nuclear reactor.

Core: That portion of a nuclear reactor containing the fuel elements.

Criticality: Term used in reactor physics to describe the state in which the number of neutrons released by the fission process is exactly balanced by the neutrons being absorbed and escaping the reactor core. At criticality, the nuclear fission chain reaction is self-sustaining.

Decay heat: Heat produced by the decay of radioactive isotopes contained in nuclear fuel.

Decay, radioactive: Disintegration of the nucleus of an unstable element by the spontaneous emission of charged particles (alpha, beta, positron) or photons of energy (gamma radiation) from the nucleus, spontaneous fission, or electron capture.

Depleted uranium: Uranium enriched in the element uranium-238 relative to uranium-235 compared to that usually found in nature. Also, uranium in which the uranium-235 content has been reduced through a physical process.

Design basis phenomena: Earthquakes, tornadoes, hurricanes, floods, and other events that a nuclear facility must be designed and built to withstand without loss of systems, structures, and components necessary to ensure public health and safety.

Design basis threat: In the context of this study, hypothetical ground assault threat against a commercial nuclear power plant. Some generic elements of the design basis threat are described in Title 10, Section 73.1(a) of the Code of Federal Regulations (10 CFR 73.1(a)).

Dirty bomb: See *Radiological Dispersal Device*.

Dry storage: Out-of-water storage of spent nuclear fuel in heavily shielded casks.

Drywell: The containment structure enclosing a boiling water nuclear reactor vessel. The drywell is connected to a pressure suppression system and provides a barrier to the release of radioactive material to the environment under accident conditions.

Dual-purpose cask: See *Cask*.

Fissile material: Material that undergoes fission from thermal (slow) neutrons. Although sometimes used as a synonym for fissionable material, the term "fissile" has acquired this more restricted meaning in nuclear reactor technology. The three primary fissile materials are uranium-233, uranium-235, and plutonium-239.

Fission: Splitting of a nucleus into at least two nuclei accompanied by the release of neutrons and a relatively large amount of energy.

Fissionable: Material that is capable of undergoing fission from fast neutrons.

Fission products: Nuclei resulting from the fission of elements such as uranium.

Fuel assembly: A square array of fuel rods.

Fuel pellet: A small cylinder of uranium usually in a ceramic form (uranium dioxide, UO_2), typically measuring about 0.4 to 0.65 inches (1.0 to 1.65 centimeters) tall and about 0.3 to 0.5 inch (0.8 to 1.25 centimeters) in diameter.

Fuel reprocessing: Chemical processing of reactor fuel to separate the unused fissionable material (uranium and plutonium) from waste material.

Fuel rod: Sometimes referred to as a *fuel element* or *fuel pin*. A long, slender tube that holds the uranium fuel pellets. Fuel rods are assembled into bundles called *fuel assemblies*.

Gamma ray: Electromagnetic radiation (high-energy photons) emitted from certain radioactive isotopes when they undergo radioactive decay.

Half-life (radioactive): Time required for half the atoms of a radioactive substance to undergo radioactive decay. Each radioactive isotope has a unique half-life. For example, cesium-137 decays with a half-life of 30.2 years, and plutonium-239 decays with a half-life of 24,065 years.

Independent Spent Fuel Storage Installation (ISFSI): A facility for storing spent fuel in wet pools or dry casks as defined in Title 10, Part 72 of the Code of Federal Regulations.

Irradiation: Process of exposing material to radiation, for example, the exposure of nuclear fuel in the reactor core to neutrons.

Isotope: Elements that have the same number of protons but different numbers of neutrons. For example, uranium-235 and uranium-238 are different isotopes of the element uranium.

Loss-of-pool-coolant event: A postulated accidental or malevolent event that results in a loss of the water coolant from a spent fuel pool at a rate in excess of the capability of the water makeup system to restore it.

Megawatt: One million watts.

MELCOR: A computer code developed by Sandia National Laboratories for use in analyzing severe reactor core accidents. The code has been adapted to model fluid flow, heat transfer, fuel cladding oxidation kinetics, and fission product release phenomena associated with spent fuel assemblies in spent fuel pools in loss-of-pool-coolant events.

Metric ton: Weight unit corresponding to 1000 kg or approximately 2200 pounds.

Metric tons of uranium: See *MTU*.

Moderator: Material, such as ordinary water, heavy water, or graphite, used in a reactor to slow down high-energy neutrons.

MTU (metric tons of uranium): Unit of measurement of the mass for spent nuclear fuel, also expressed in metric tons of heavy metal (MTHM). It refers to the initial mass of uranium that is contained in a fuel assembly. It does not include the mass of fuel cladding (zirconium alloy) or the oxygen in the fuel compound.

Multi-purpose cask: See *Cask*.

MWe: Megawatts of electrical energy output from a power plant.

MWt: Megawatts of thermal energy output from a power plant.

Neutron: Uncharged subatomic particle contained in the nucleus of an atom. Neutrons are emitted from the nucleus during the fission process.

Open rack: A storage rack in a spent fuel pool that has open space and lateral channels between the cells for storing spent fuel assemblies to permit water circulation.

Overpack: Metal or concrete cask used for storage or transportation of a canister containing spent nuclear fuel. See Cask.

Owner-controlled area: That part of the power plant site over which the plant operator exercises control. This usually corresponds to the boundary of the site.

Pellet: See *Fuel pellet*.

Penetrate: To pass into, but not completely through, a solid object.

Perforate: To produce a hole that goes completely through a solid object.

Plutonium-239: A fissile isotope of plutonium that contains 94 protons and 145 neutrons.

Pressurized water reactor (PWR): A type of nuclear reactor in which the reactor's water coolant is kept at high pressure to prevent it from boiling. The coolant transfers its heat to a secondary water system that boils into steam to drive the turbine and generator to produce electricity.

Probabilistic risk assessment: A systematic, quantitative method to assess risk (see below) as it relates to the performance of a complex system.

Protected area: A zone located within the owner-controlled area of a commercial nuclear power plant site in which access is restricted using guards, fences, and other barriers.

psia: Unit of pressure, pounds per square inch absolute, that is the total pressure including the pressure of the atmosphere.

Radioactivity: Spontaneous transformation of an unstable atom, often resulting in the emission of particles (alpha and beta) or gamma radiation. The process is referred to as radioactive decay.

Radiological Dispersal Device (RDD): A terrorist device in which sources of radioactive material are dispersed by explosives or other means. Also referred to as a *dirty bomb*.

Radiological sabotage: Any deliberate act directed against a nuclear power plant or spent fuel in storage or transport that could directly or indirectly endanger the public health and safety by exposure to radiation.

Radionuclide: Any form of an isotope of an element that is radioactive.

Re-racking: Replacement of the existing racks in a spent fuel pool with new racks that increase the number of spent fuel assemblies that can be stored.

Risk: The potential for an adverse effect from an accident or terrorist attack. This potential can be estimated quantitatively if answers to the following three questions can be obtained: (1) What can go wrong? (2) How likely is it? (3) What are the consequences?

Safety: In the context of spent fuel storage, measures that protect storage facilities against failure, damage, human error, or other accidents that would disperse radioactivity in the environment.

Safeguards: As used in the regulation of domestic nuclear facilities and materials, the use of material control and accounting programs to verify that all nuclear material is properly controlled and accounted for, and also the use of physical protection equipment and security forces to protect such material.

Safeguards information: Information not otherwise classified as National Security Information or Restricted Data that specifically identifies a U.S. Nuclear Regulatory Commission licensee's or applicant's detailed (1) security measures for the physical protection of special nuclear material or (2) security measures for the physical protection and location of certain plant equipment vital to the safety of production or utilization facilities (10 CFR 73.2). The U.S. Nuclear Regulatory Commission has the authority to determine whether information is "safeguards information."

Security: In the context of spent fuel storage, measures to protect storage facilities against sabotage, attacks, or theft.

Shaped charge: A demolition and wall penetration or perforation device that uses high explosive to create a high-velocity jet of material.

Single-purpose cask: See *Cask*.

Special nuclear material: Fissile elements such as uranium and plutonium.

Spent fuel: See *Spent nuclear fuel*.

Spent fuel pool: A water-filled pool that is used at all commercial nuclear reactors for storage of spent (used) fuel elements after their removal from a nuclear reactor. Spent fuel pools are constructed of reinforced concrete and lined with stainless steel. The inside of the pool has storage racks to hold the spent fuel assemblies and may contain a gated compartment to hold a spent fuel cask while it is being loaded and sealed.

Spent (or used or irradiated fuel) nuclear fuel: Fuel that has been "burned" in the core of a nuclear reactor and is no longer efficient for producing electricity. After discharge from a reactor, spent fuel is stored in water-filled pools (see *Wet storage*) for shielding and cooling.

Storage-only cask: See *Cask*.

Thermal power: Total heat output from the core of a nuclear reactor.

Uranium-235: A fissile isotope of uranium that contains 92 protons and 143 neutrons. It is the principal nuclear fuel in nuclear power reactors.

Uranium-238: An isotope of uranium that contains 92 protons and 146 neutrons.

Vital area: A zone located within the protected area of a commercial nuclear power plant site that contains the reactor control room, the reactor core, support buildings, and the spent fuel pool. It is the most carefully controlled and guarded part of the plant site.

Watt: Unit of power.

Watt-hour: Energy unit of measure equal to one watt of power supplied for one hour.

Wet storage: Storage of spent nuclear fuel in spent fuel pools.

Zircaloy: Zirconium alloy used as cladding for uranium oxide fuel pellets in reactor fuel assemblies.

Zirconium cladding fire: A self-sustaining, exothermic reaction caused by rapid oxidation of zirconium fuel cladding (zircaloy) at high temperatures.

F

ACRONYMS

ACRS: Advisory Committee on Reactor Safeguards

BAM: Bundesanstalt für Materialforschung und -prüfung

BMU: Bundesministerium für Umwelt, Naturschutz und Reaktorsicherheit

BNL: Brookhaven National Laboratory

BWR: Boiling Water Nuclear Reactor (see Appendix E)

CFD: Computational Fluid Dynamics

DBT: Design Basis Threat (see Appendix E)

DHS: United States Department of Homeland Security

DOE: United States Department of Energy

EPRI: Formerly referred to as the Electric Power Research Institute

GAO: United States Government Accountability Office (formerly the General Accounting Office)

GESMO: Final Generic Environmental Statement on the Use of Recycled Plutonium in Mixed Oxide Fuel in Light-Water Cooled Reactors

GNB: Gesellschaft für Nuklear-Behälter, mbH

GNS: Gesellschaft für Nuklear-Service, mbH

GNSI: General Nuclear Systems, Inc.

GRS: Gesellschaft für Anlagen- und Reaktorsicherheit, mbH

GWd/MTU: Gigawatt-Days per Metric Ton of Uranium (see *Burn-up* in Appendix E)

INL: Idaho National Laboratory (formerly Idaho National Engineering and Environmental Laboratory)

ISFSI: Independent Spent Fuel Storage Installation

HSK: Die Hauptabteilung für die Sicherheit der Kernanlagen

MTU: Metric Tons of Uranium (see Appendix E)

MWd/MTU: Megawatt-Days per Metric Ton of Uranium (see *Burn-up* in Appendix E)

NPP: Nuclear Power Plant

NRC: National Research Council

PFS: Private Fuel Storage

PWR: Pressurized Water Nuclear Reactor (see Appendix E)

RDD: Radiological Dispersal Device (see Appendix E)

RPG: Rocket-Propelled Grenade

RSK: Reaktorsicherheitskommission

TOW: Tube-Launched, Optically Tracked, Wire Guided [Missile] (see Appendix E)

USNRC: United States Nuclear Regulatory Commission